The River Wolf

The River Wolf

KEITH AND LIZ LAIDLER

London
GEORGE ALLEN & UNWIN
Boston   Sydney

George Allen & Unwin (Publishers) Ltd,
40 Museum Street, London WC1A 1LU, UK

George Allen & Unwin (Publishers) Ltd,
Park Lane, Hemel Hempstead, Herts HP2 4TE, UK

Allen & Unwin Inc.,
9 Winchester Terrace, Winchester, Mass 01890, USA

George Allen & Unwin Australia Pty Ltd,
8 Napier Street, North Sydney, NSW 2060, Australia

First published in 1983

British Library Cataloguing in Publication Data

Laidler, Keith
    The river wolf.
1. Otters
I. Title
599.74'447   QL737.C25

ISBN 0-04-599008-5

Set in 10 on 11 point Baskerville by Inforum Ltd, Portsmouth
and printed in Great Britain
by Mackays of Chatham

Contents

Illustrations

PHOTOGRAPHS

Between pages 82 and 83

Russel Lake, east canal
The broad-fronted caiman
Katina with his trophy
The manatee
Lama Creek
The ocelot
Three-toed sloth
Marking site
Close-up of latrine
Freshly deposited scent
The Streaky group periscoping
Mister and Missus
A streak of silver
Resting in between hunting bouts
Hunting in the shallows
Feeding in the shallows
Landing a fish
Excavating a signal spot
Sleeping
Grooming
Bluff-charging the canoe
Giant otter skins and seller
The fangs of a *fer de lance*
Niblet
Nibs at play
Niblet gorging on a meal of patois
A favourite oasis
A weekly washday hindrance
Waiting to be chased out of the fire bucket
Wobble-belly
The metronome
Feet are never safe with an otter around

Saved by Keith
The day of Nibs' release

DIAGRAMS

Acknowledgements

We would like to thank the Ministry of Agriculture, the Ministry of the Interior and the National Science Research Council of Guyana for granting permission for the study. Thanks also to Professor J. J. Niles, Mr Adrian Thompson, Mr Stanley Lee and Mr Burnham for their support in the building of Squeaky's enclosure at Georgetown Zoo; to the Chairman of the East Demerara Water Conservancy, Mr Chandra, for allowing us to use the rest-house at Lama; to the British High Commissioner, Mr Philip Mallet, and his wife, Mary, for suggesting Lama to us in the first place; to Lawrence and Cleo Van Sertima for their easy hospitality and the help they gave us in finding food supplies at critical times; and to Deodat and Wally John for guiding us through the treacherous expanse of Bonsica Lake.

The expedition was partly financed by the Otter Trust, the Fauna and Flora Preservation Society, the World Expeditionary Association and the Caribbean Conservation Association.

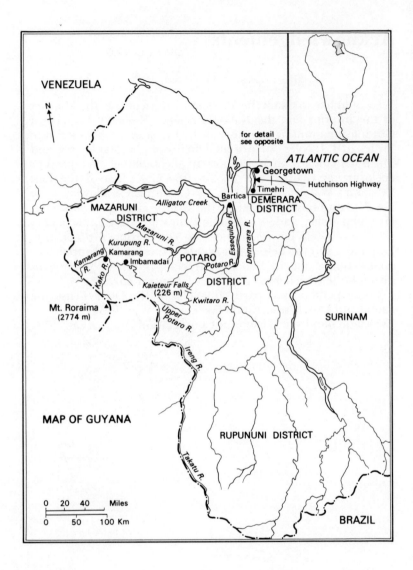

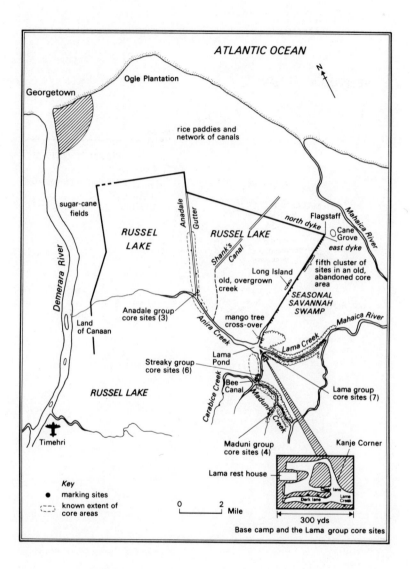

ATLANTIC OCEAN

Georgetown

Ogle Plantation

rice paddies and
network of canals

sugar-cane
fields

Anadale
Gutter

RUSSEL
LAKE

RUSSEL LAKE

north dyke

Flagstaff

Cane
Grove

east dyke

Mahaica River

Demerara River

Shank's
Canal

Long Island
old, overgrown
creek

fifth cluster of
sites in an old,
abandoned core
area

SEASONAL
SAVANNAH
SWAMP

Land
of Canaan

Anadale group
core sites (3)

Anira Creek

mango tree
cross-over

Lama Creek

Mahaica River

Streaky group
core sites (6)

Lama
Pond

RUSSEL LAKE

Carabice Creek

Bee
Canal

Madumi Creek

Lama group
core sites (7)

Timehri

Maduni group
core sites (4)

Lama rest house

Kanje Corner

Key
● marking sites
- - - known extent of
core areas

0   2
|_____| Mile

Clear lane
Dark lane
Lama
Creek

300 yds
Base camp and the Lama group core sites

1   A Streak of Silver

(Keith)

They were swimming all round the canoe, red mouths ablaze, barking angrily and getting closer every second. After four days of searching along the black water creeks of Guyana, this was our first view of *Pteronura brasiliensis*, the giant otter, and from where I was sitting in the front of a native canoe, I thought it might very well be our last. 'Scientist and Cameraman Eaten Alive in Guyana': I could see the headline in *The Times* sandwiched between the increase in the price of baked beans and the results of the 3.45 at Ascot.

It was no idle speculation; as its name suggests, the giant otter is a formidable creature. More than twice the size of the European or North American varieties (lengths in excess of seven feet have been reported), this tropical otter can reach seventy pounds or more. The animals now busily surrounding our canoe were no record-breakers, but they were certainly not flyweights either. Nor were they entirely harmless; the giant otter feeds mainly on fish and boasts a redoubtable array of needle-sharp teeth for dealing efficiently with its slippery prey. Teeth that can zip through the tough scales of a two-foot fish with the ease of a sewing machine through velvet would make short work of human flesh. In 1969 a keeper at São Paulo Zoo in Brazil accidentally fell into an enclosure containing two giant otter cubs which were jealously defended by their parents. The unfortunate man was swiftly pounced upon by the adult otters and literally chewed to death. And the worst of it was that the bellicose otters now busily surrounding our canoe were also a family group, with the parents in the vanguard of the attack as if protecting their offspring. I looked behind at Liz who was steering while I filmed. She was obviously as worried as I was.

It had all begun far more auspiciously. We had set off from our rest-house base just after dawn, in the cool first hour of morning before the sun reaches over the top of the trees and sends the temperature soaring into the nineties. We had by this time mastered the use of the long, knife-shaped Amerindian paddles and moved steadily and quietly under the overhanging archway of trees that lined the first hundred yards of waterway. The first bend in the creek, where the trees receded as the river widened, we had already named Kanje Corner in honour of its permanent residents, the kanje pheasants. These birds, better known as hoatzins (and less kindly as stinking nannas), are about as close as one can come today to the primitive lizard-birds of 150 million years ago. The adult bird is slightly larger than an English pheasant, with brown-black plumage and a large blue head-crest, making it look like a cross between a partridge and a cockatoo. Many characteristics distinguish the hoatzin from other birds, so much so that zoologists have placed it in a family and sub-order of its own. It does not sing or call but instead gives voice to a bronchitic wheeze, a trait which made us look rather foolish during our second day on the river: on first hearing these wheezings, we took the sound to be the breathing of an otter and moved in close to investigate. As the birds spend most of their time deep in the dense bush at the sides of the creek, we waited over an hour for the 'otter' to emerge before realising our mistake.

The hoatzins of Kanje Corner heralded our every arrival with a chorus of asthmatic gaspings, but the loudest cacophony came from what, pound for pound, is certainly one of the noisiest creatures on God's earth – the howler monkey. Just after dawn and at odd times throughout the day, a group of howlers that 'owned' the stretch of riverine forest just downriver from the rest-house would begin their clamorous chanting. Starting as a pumping growl that grew rapidly into a series of guttural, booming roars, their calls were guaranteed to wake one from sleep far more effectively than any alarm clock. The males have by far the stronger call and their vocal apparatus has been greatly modified by evolution for this task. The sound originates in the larynx, but it is a specially enlarged, bony sound-box – the hyoid apparatus – that greatly amplifies the original sound. The howler's call can be heard at distances of two miles or more and to hear a troop from less than a hundred yards is an ear-ringing experience. Their calls serve to tell neighbouring groups that an area is occupied and that the residents are willing and able to fight to keep possession. In many ways it's a very civilised way to conduct 'warfare' – vocal battles are a lot less damaging to both sides than physical combat, a fact *Homo sapiens* would do well to learn.

The Guyanese, who know these monkeys well and sometimes hunt them for the pot, have a much more pleasing explanation for the male howler's calls. They say that the male produces such stentorian tones because the female, angry with him for some real or imagined misdemeanour, takes his testicles between her palms and rubs them together – hard. Any human male will, I think, vouch for the intuitive 'rightness' of the Guyanese explanation.

We ran the vocal gauntlet of howler calls on a long straight stretch of river. Some time later, rounding the fourth bend of Lama Creek, we paddled straight into our first otter group. The first sign of their presence was a streak of silver on the gun-metal surface of the water about 300 yards away. We did not realise the significance of this at first, as fish were often breaking surface after flies and other insect life. But this time the silver streak was bigger and, unlike those of fish, it persisted in the water for a long time. Puzzled, we scanned the area with binoculars and telephoto, and through them we could see four grey-brown heads moving swiftly through the water, with the occasional curved outline of a back as one of the otters dived. Every so often one of them would surface with a gleaming fish and swim leisurely to the shallows to consume its prize. They were otters all right, but were they giant otters? From that distance it was hard to judge size, and imposs-ible to say whether the animals were of the smaller *Lutra enudris* species or our much-sought *Pteronura*. Slowly, hardly moving her paddle, Liz took us closer and closer to the otters, who were still completely oblivious of our presence. So intent were they on procuring a hearty breakfast that we were able to come within forty feet of the family group without being noticed. Then one, the big male who must have measured all of seven feet, surfaced near us after a successful sortie beneath the lily pads on the left-hand side of the creek. If ever an otter looked surprised, he did. He obviously never expected to find humans so close at this time in the morning for he gave a deep, coughing snort of warning – losing his hard-earned fish in the process – and dived again beneath the brown-black waters of the creek. The rest of the family disappeared a half-second later in a flurry of dark foam.

But the close-up view had been enough. We now knew for certain that these creatures were the species we were looking for. We had found our giants. 'Hello, goodbye', I thought ruefully, scanning the waters of the now-empty creek. We had, I felt, seen the last of our otter family, at least for today. Over the past eight years we had studied and filmed a wide variety of wild animals in many of the remoter regions of the world. Everything from manacou to monkey, from parrot to pangolin, had come under our scrutiny in habitats as diverse as the West African jungle and the arid deserts of Oman. In every case the wild creature's re-sponse was the same – as soon as Man appeared on the scene the

3

animal would attempt to put as much space as possible between itself and the arch-hunter. I had no doubt that the giant otter would react in precisely the same way.

I was wrong. I soon discovered that the giant otter had very little fear of anything, and certainly not of Man. Our group resurfaced together about thirty yards from our boat, sheltering behind the long, rough stems of a stand of moca-moca, that over-sized lily that lines the banks of most rivers in tropical South America. Here they conferred for a while in short squeaks, squeals and grunts before launching their quite terrifying counter-attack. The big male came first, followed by the female, each of them swimming a zig-zag path towards us, their open mouths bright red against the deep purple-brown of their coats. We simply had not expected this reaction and just sat quietly in the boat, excitement and fear mounting with each of their sallies towards us. Behind them the two cubs began their own meandering attack in imitation of their parents. Then the barking began: a deep, growling, throaty bark bellowed from their throats as the two parents raised themselves vertically in the water so that the top half of their long bodies stood clear of the creek. The bib of yellow-white fur down the front of their chests was clearly visible and I noted with satisfaction that each bib had a different shape – it seemed that we would be able to identify the otters individually, an important aid in unravelling the intricacies of their social life.

They were now less than twenty-five feet from the front of the boat, still barking and apparently very angry at our intrusion. I could see their pointed, needle-like teeth clearly, and the idea of providing fodder for that evil-looking maw was not appealing. Yet despite my anxiety I could not help admiring the otters for their boldness. After all, this river was their home and they had every right to see off intruders. But we also had a right to protect ourselves and it looked increasingly as if we might have to exercise that privilege. They were by this time only fifteen feet away and I had long since exchanged my camera for a razor-sharp cutlass, ready to strike out should they attempt to come aboard the canoe. Then suddenly, and as if by command, the four otters dived and disappeared.

We waited, silent and expectant, for their re-appearance, but we neither saw nor heard anything. It was as if the otters had been wraiths, insubstantial dream animals, they had faded away so successfully and completely. There was only the lapping of the creek waters, the wind soughing through the topmost branches of the forest and, above it all, the toc-toc-tocoo of Lama Creek's resident toucan. After fifteen fruitless minutes we turned the canoe and began the long, three-mile paddle back to the rest-house.

The river was quiet once more, but my mind was in a turmoil. Our first sighting of giant otter had been quite fabulous, but their aggressive behaviour did pose problems for the way in which we had planned to carry out both the filming and the research aspects of the expedition. Neither of us had expected the otters' response to be quite so violent, rather the opposite. We imagined them shy creatures, and had planned to overcome their fear gradually by acting calmly and quietly whenever we were near them, and so to accustom them to our presence. But now we were faced with exactly the opposite dilemma. We now had to behave in such a way that we acted less as a red rag to a bull with the otters, so that we gradually became part of the furniture of the forest and were ignored. If we could achieve that, many of the mysteries surrounding this unique creature might be solved.

We had been on the river for more than four hours by now, and chronic 'Lama-bum' was beginning to set in. This condition first manifests itself as total, icy numbing of the nether regions, caused by enforced squatting on the hard, narrow planks which imper-sonate seats in a *kurial*, or native canoe. The affliction progresses to a dull ache across one's bottom and rapidly escalates into a burning agony reminiscent of my school-day canings. The only remedy known to Man, and the one that I now utilised with some relief, is to stop paddling, stand and rapidly massage the tender portions. The morning after our first day on the river had been worst – we had eaten breakfast standing.

Thoughts of nice hot food caressed my mind as we paddled back, now that we had persuaded Pappy, the rest-house cook, not to grace us with his Guyanese specialities. Pappy was short and rather fat, with a winning smile and huge, owl-like eyes. Like most West Indians, he took life's problems and deadlines at a dead crawl. He could however stir himself when occasion demanded and so it had been on our first night at the rest-house. We had arrived home late after a gruelling day on the river to find that Pappy had 'a good t'ing cookin' '. The good thing had a quite excellent aroma and turned out to be a cauldron full of the most delicious-looking stew. Despite our questions, Pappy refused to be drawn on the contents of the stewpot. 'You try he,' he answered, placing the pot on the camp table and smiling secretly. 'Mek all pepple strong-strong.' That was the only explanation we could get, so Liz set to with the ladle. She pulled out a heap of meat, vegetables, peas and lentils that looked positively mouth-watering. My turn. In went the ladle and out came . . . I could hardly believe my eyes: sticking out from among the peas and lentils was a human hand. A baby's hand!

'You got he!' Pappy cried out in delight. 'One more to meet!' he continued excitedly, 'but dis one de mistresses'.' He wagged an admonishing finger in my direction. 'You can't have bot'.'

I tried to remain calm and to talk quietly to our cannibal-cook. 'Pappy,' I said, poking at the hand with feigned nonchalance, 'er, what is this?'

'Best part o' de stew,' Pappy enthused. 'It sweet meat, mek strong-strong. Monkey hand and lentils!'

While we sat in shocked silence, Pappy went on to explain just why this delicacy was so prized. He had shot the monkey – a capuchin – just after we had left the rest-house. Once the corpse was skinned and gutted, the two hands were placed in the pot along with the pulses and other ingredients. A fire was lighted under the cauldron, and the heat caused the hands to clench tightly, trapping the peas and lentils in a dead-man's-grip. The pulses cook in the juices of the monkey hand, and the whole macabre ensemble is highly regarded in Guyanese culinary circles. Needless to say, these esoteric pleasures were somewhat lost on us newcomers. Liz was grey in the face and I had suddenly lost my appetite. It's not that we are finicky about food. On our various trips we have eaten a great many unusual dishes: potted pangolin, roast iguana, ragout of cane-rat, bat stew, even the proverbial sheep's eye, have all graced our palates. But a monkey's hand. Somehow I just could not bring myself to do it; the shape and the feel of the delicacy so lovingly cooked by Pappy, reminded me irresistibly of a baby's small hand. Consumption was out.

But how to get out of our spot without offending Pappy? He had, in all good faith, spent several hours cooking his pièce de résistance as a special treat for his guests. I was nonplussed but fortunately Liz came to the rescue. With rare acting ability, she wiped her hand across her brow and remarked how hot it had become, hadn't it? Yes, yes, I echoed, unsure of her plan but glad of any chance, however forlorn, of avoiding our meal, yes, it had become hot and, indeed, very still too. Perhaps then, said Liz, fixing me with an unblinking stare, we might take our delicious meal out onto the jetty, where we could benefit from the riverside breeze.

Pappy was looking extremely puzzled at this turn in the conversation, but as I knew he already thought anyone who would look for otters without shooting them slightly insane, I was sure he would regard this strange notion as yet another confirmation of our eccentricity. What a jolly good idea, I told Liz heartily (Pappy was an aficionado of P. G. Wodehouse and expected all eccentric Englishmen to say 'jolly good'), and we rapidly repaired to our jetty-side picnic area. The ants and mosquitoes were numerous and uncomfortable, but it was well worth the inconvenience. Monkey hands and most other parts of the stew were quickly consigned to the creek waters, to the obvious delight of the red piranha who swirled around the jetty in their eagerness to partake of the feast. That night we had gone to bed hungry, after insisting to Pappy that he should not tire himself unduly by cooking such superb specialities. Far better for him to conserve his creative energies by cooking up the dehydrated beef stews and chickens curries we had brought with us from England. His culinary masterpieces should not be squandered on everyday meals. He should, we told him with something approaching panic, save his unique talents for special cocasions. Like when we left for England.

I had been so engrossed by thoughts of food that I failed to take account of an ominous humming sound coming from one side of the river. Not until a black cloud of insects broke from the riverside vegetation and came skimming towards us about fifteen feet from the top of the water did I realise what we were facing. Killer bees!

These insects are the result of a breeding experiment that went horribly wrong. The European bee, *Apis mellifica*, is famous for the huge quantity of honey it produces in temperate climes. Unfortunately, it does not do well in tropical South America. By contrast, the native bee, being perfectly adapted to its hot and humid habitat, thrives in the jungle environment. But it has one serious drawback: the native bee does not produce the vast quantities of honey that would make it a commercial proposition as far as Man is concerned. So Man took the problem in hand and began crossing different strains of bees from different parts of the world in an attempt to produce the perfect South American bee, a high-yield variety that would thrive in the taxing climate of the tropics.

One experiment crossed a native bee with a strain of African bee. The result was disastrous. The new bee was perfectly adapted to the South American climate but, besides being a high-yield variety, the offspring of this unnatural union proved ultra-aggressive. Simple proximity was enough to make any animal – including Man – a target, and once one bee stung, the rest of the colony were quick to follow. Had Man been in control, as he always claims to be in experiments of this kind, this

7

particular breed should have been swiftly exterminated. Unfortunately, they escaped from the breeding station and proved so well adapted that killer bee colonies multiplied rapidly, spreading westwards and northwards into continental South America. Despite strenuous attempts at eradication it is still adding to its range, pushing inexorably towards Mexico and the United States border. According to one expert, by 1984 the killer bee will have reached Texas. And as its range increases, so does the list of fatalities. During the first two days of our visit to Guyana we saw a man, horribly swollen, being removed from the Botanic Gardens in Georgetown, Guyana's capital city. He had been walking quietly in the gardens and had unknowingly strolled beneath a tree containing a swarm of killer bees. True to form, the bees had fallen on him without warning, covering his body with stings. We discovered later that he had died in hospital.

There was nothing we could do as the bee cloud approached. We simply sat as still as we could, hardly daring to breathe, and hoped that the sinister swarm would fly over without troubling us. The menacing buzz grew in intensity until it blocked out all other sounds. Then it began to recede and I realised with relief that we were through the worst. But a straggler must somehow have taken exception to me because the next thing I knew was a searing pain in my left shoulder. I looked down in alarm and saw six more bees clustering chummily around my left bicep. Almost instantly, this area went white-hot as more stings thrust into my flesh. That was enough for me. 'Into the water!' I yelled to Liz, and prepared to fling myself overboard. There were piranha and caiman in the river, but at that moment there was infinitely more chance of dying above water than below it. I heard Liz splash into the river behind me and I followed her a second later, though not before other bees had bestowed their parting gifts on my arm, shoulder and ribs.

I surfaced with relief, glad to have escaped in one piece from the small black marauders. But I had only just taken my first breath when I felt again the burning pain of a bee sting, this time on my nose. The bees were still there! As I submerged, I recalled with horror the tale a Guyanese farmer had recounted three days earlier. His brother had fallen foul of the bees while weeding on the side of a dyke and, like us, had taken refuge in the water. The bees had hovered above him too, rushing in to attack each time he emerged from the water to breathe. By the time they tired of this sport he was severely stung about the hands and face and almost unconscious. Was this to be our fate?

This time I stayed under water until my lungs felt like two

red-hot coals inside my chest, praying fervently that the bees would move on. Then I broke surface as quietly as I could and tried to grab a quick chestful of oxygen. But once again the bees were faster and I submerged with two additional stings, one on my right hand and another on my ear. Things looked bad for me and I wondered how Liz was faring. The next time I emerged unwillingly from the water, my head struck against something soft. It was Liz's wide-brimmed straw hat. Almost without thinking, I seized the hat and flung it over my face. Through the plaited raffia I saw four killer bees smash into its crown, under which I could, by treading water, at last breathe safely.

'Liz!' I gasped desperately, 'Liz! You OK?'

'Almost.' It was a whisper, and she didn't sound as breathless as I did. I realised with relief that she must be relatively unharmed.

I was still treading water madly, and now that I was in no immediate danger from the bees, thoughts of unpleasant meetings with piranha were beginning to surface in my mind. 'Liz? Where are you?'

'At the river bank, hiding under a tree. I swam here underwater as soon as they attacked. Only three stings.'

I have never been very proficient at treading water, and by this time I felt thoroughly exhausted, though still safe thanks to Liz's blessed hat. 'Any room?' I asked hopefully. My left arm was difficult to move and felt enormously swollen.

'Plenty,' came the happy reply. 'If you turn right and swim underwater you'll hit bottom after about twelve feet. Don't bother bringing the hat, leave it there.'

I had no intention of bringing the damned thing but I was too exhausted to answer. Instead, I took a good lungful of air, sank once more beneath the water and, abandoning the safety of the straw hat, struck out for the bank. Sure enough, after several tense seconds, I felt the mud, leaves and dead branches of the sloping riverside. I dreaded resurfacing – had the bees followed? But I was running out of air. Up I went, slowly and softly.

No bees. Just cool, dark air under a blanket of green leaves that hung down to the water line. Liz was about five feet away; I could just make out two large red swellings on her right cheek, but apart from that she seemed fine. I couldn't see the bees but I could certainly hear them. They hung around the boat for another four or five minutes before the angry buzzing diminished to nothing and we were left listening to the other, less menacing sounds of creek life.

After a further ten minutes I peered cautiously from between the branches. Nothing. We collected the canoe, or rather Liz did, for my left arm was now so swollen that it was virtually useless, and I was in considerable pain from what Liz estimated as a total of seventeen stings. Despite that, we rowed away from the safety of the rest-house at first. I felt I had to collect the straw hat. After it had probably saved my life, I just could not bring myself to leave it behind.

I was feeling so groggy when we finally berthed the canoe that I spent the rest of the day in my hammock, pumped full of pills and anti-histamine. Lying there with both my left arm and my right temple throbbing angrily, I longed for the end of the trip, for our little Durham cottage and for the safe land of Britain. Oh for some proper food, cooler weather and a good night's sleep on a real bed with no ants, mosquitoes, centipedes or snakes to disturb one's slumbers! I found it difficult to conceive how I could ever delude myself into the belief that these wretched trips were fun. It was even harder to believe that our present troubles owed themselves entirely to a spilt drink in a West End wine bar.

2 The River Wolf

(Keith)

I was bored, Liz was bored. After two years in London, the delights of the city were beginning to pall. Sitting at a cramped table in our favourite Kensington wine bar, squashed between a quiet, self-absorbed negro on one side and two strident students earnestly discussing the sociological relevance of Marxist philosophy to the struggle for Women's Liberation on the other, we felt that the capital city had finally lost the golden lustre that had so attracted us when we arrived, raw provincials, twenty-four months earlier. Then it was all bright lights, the theatre, art galleries, shows, museums; now it was grime, dirt and the knowledge that Kensington Gardens, green though it might be, was an illusion, a tiny island of vegetation in 400 square miles of concrete. We longed for green fields, country lanes, cows, crows, rabbits, the whole rural bit.

I had spent the time in London working as a scriptwriter for Survival, the world-renowned television wildlife company. It was exciting work. I would 'adopt' an animal, reading all I could about its life cycle, behaviour, anatomy etc for the two months it took to turn 10,000 feet of raw footage on the beast into a cogent, 918-feet, 25-minute story detailing the most important aspects of its life. I learned far more about animal behaviour – the aspect of zoology that had always appealed to me – in Survival's prestigious Park Lane offices than I had ever been taught at University. To my academic friends it seemed the most ideal of jobs, and I continually received letters which began 'Hope this finds you in the country . . .', 'Hail, Jet-setter . . .' or 'Remember me on your next world cruise . . .'. They seemed to believe that I visited every exotic place I wrote about, skimming off by plane across half the world to absorb the essential atmosphere of the place, to watch the animal first hand, before I set pen to paper. The truth was more prosaic. The only things I skimmed across were the pages of the many hundreds of books in Survival's reference library. From there, I would pack my bags and travel to my basement office, or perhaps take a holiday and pop along the corridor to drop in on a film editor.

I was, in a word, desk-bound. Among the notices stuck to my office door was one which read 'Beware – Cape Hunting Dog'. Most people were puzzled, some thought it a joke, but in fact I was in deadly earnest. The Cape hunting dog feeds its young by regurgitating food. That was exactly how I felt: I digested innumerable boring papers written mostly by long-winded, jargon-filled, pompous academics, and then regurgitated them in digestible form to my young – the viewing public. It was a dog's life, and just to rub salt in the wound, every so often a fit, bronzed cameraman would stride past my office, fresh from some tropical paradise where he had filmed the unusual and interesting animal I would write about in my next half-hour programme. Suddenly, it was not enough to read or write about the animals. That was living one's life second-hand. I wanted to see the beasts in the flesh, and the best way to do that was to become a cameraman. But how? And what, and where, would I film? Africa, the photographer's happy hunting ground, probably had more photographers per square mile than any other large mammal. Where could we go?

Liz was if anything even worse off. She was working as a research assistant at the Royal Free Hospital School of Medicine's pharmacology laboratories, dosing cagefuls of guinea-pigs with new drugs and monitoring the effects, if any, which the chemicals had on the unfortunate creatures' internal organs. It was worthy research but soul-destroying work for a girl whose greatest desire in life was to study animals in their natural habitat, describing their behaviour and unravelling the mysteries of their social life. But Liz had much the same problem as me: which animal should she choose to study, and in what part of the world would the research take place?

So, as we sat in the wine bar and sipped our Liebfraumilch, we looked back with nostalgia on our student days when we had organised and led two expeditions to the rain forests of St Vincent to study the rare and endangered St Vincent parrot, or when we had journeyed to a village in Ghana where the monkeys of the surrounding forest were considered 'children of the gods' and were so tame that they walked the village streets like dogs. Yes, we reminisced, those were the days.

It was terrible. Here we were in our mid-twenties and mooning over our memories like a pair of pensioners. Surely we could do something to get out of the rut we found ourselves in? There was one thing I could do: I could get some more wine, our glasses were empty. I stood up. Unfortunately, the two Women's Lib students chose to rise at the same instant. I rebounded off the nearest of them, cannoning into the pensive negro on our right and spilling his drink. Apologies were offered and accepted and I insisted he let me buy him another drink. By the time I returned from the bar, he and Liz were deep in conversation about the relative merits of St Vincent and Barbados as tourist centres.

Liz introduced our new acquaintance as Osbert, a short, stocky black in his early thirties with massively built shoulders and, when he chose to show it, a grin like a shark. In an obviously West Indian accent which I could not place to any of the Caribbean islands, he told us he was visiting London on a three-week silvi-culture course. He was beginning to feel quite homesick and, nodding towards Barbados-born Liz, he said how glad he was to talk to someone from his own part of the world. Yes, it was too cold here, too many people, too much concrete. He longed to return to his job as a forester in the jungles of Guyana's interior.

My ears pricked at this information. One of my special interests has always been the Yeti and kindred beasts, and Guyana, I knew, had tales of one such creature, a five-foot hairy man called the Didi. Osbert flashed his shark's teeth at me when I asked him whether he had seen this South American abominable snowman. He knew of the Didi but emphatically denied its existence. In his opinion, anyone who claimed a sighting of the creature had obviously been drinking too much 'high wine', the thrice-distilled rum of Guyana. By way of compensation, though, he began to tell us that he himself had seen something almost as good as the Didi. Had we heard of the 'big water dog', the 'water tiger', the animal that zoologists call the giant otter?

Liz had always been interested in otters and she nodded vaguely, but I had to confess total ignorance of the subject. Looking substantially more pleased with himself, Osbert launched into his tale of a huge otter, 'big like jaguar', that he often saw in the Mazaruni region of Guyana. It was almost black in colour, with an enormous red mouth, and it could be very dangerous. Most impressive of all, it could be found in large packs on the creeks and rivers of the Mazaruni. Osbert claimed to have run into a group of these monsters that numbered in excess of forty animals! The number seemed incredible, and I asked jokingly if Osbert hadn't himself been under the influence of 'high wine'. Our friend indignantly insisted that he had not been drunk, and went on to tell how one of his companions had shot at and killed one of ten cubs in the otter pack. Immediately, they were charged by all the adult otters. The killer, panicked by the sight of so many red-mouthed monsters seemingly intent on murder, tried to fire into the approaching pack. His gun jammed and the otters continued their attack. Osbert was convinced that their canoe would have been overturned and everyone aboard killed had he not had the presence of mind to grab the largest cooking pot in the boat and beat his cutlass against it like a drumstick. The other members of the party did likewise, and the tumult made the otters think twice about attacking, though they hung around the boat for some time, barking and snarling. Osbert said that these creatures were quite common in some parts of Guyana, and that he had seen them in Brazil, too, where they went by the local Spanish name, lobo del rio, the river wolf.

Osbert finished his story and I slowly turned my head to look at Liz. She was staring at me, too, and I knew that the same thought was buzzing around inside both our heads. Fate seemed to have led us to Osbert and to have offered us a chance to slip the reins of boredom and do what we claimed we always wanted to do. This was the one. Even the animal's name, river wolf, seemed charged with mystery and danger. The giant otter was a natural for filming, and for study.

During the next few weeks I was again leafing through reference books, hunting down obscure journals and taking notes from long-forgotten accounts. But this time there was a difference: this time Liz and I were working for ourselves, collecting, sifting and collating information on a project that would result ultimately in our escape from routine. Or so we hoped. Liz had received favourable replies from Cambridge University's Department of Applied Biology to her application to study the giant otter, but no-one in the TV world yet knew of our plans to film the beast. And once they did know, there was no guarantee that anyone would accept us, completely untried as we were in the competitive world of wildlife photography, or help to fund our trip. All we could do was to present as full an account as possible of the giant otter, to show what a fascinating beast it was, to explain how we would go about filming it, and to hope that someone would fund the work. Liz was making a parallel effort with academic funding bodies to raise cash for her study.

One thing that we learned almost immediately was that very little was known about the habits of the giant otter. This was good for Liz's fund-raising efforts – no-one will finance work on an animal whose behaviour is already well documented – but it made my job more difficult. Luckily, other aspects of the animal were just as exciting.

As its name suggests, the river wolf, known to science as *Pteronura brasiliensis*, is the largest of the world's seventeen species of otter. Measuring up to seven feet in length, the giant otter is about

14

half as long again as any other otter species and twice as long as both the European otter, *Lutra lutra*, and its North American counterpart, *L. canadensis*. South America is something of an otter paradise: it boasts more species of otter than any other continent. Apart from the giant otter, there is the little-known sea otter, *Lutra felina*, who lives out its life along the sea coast of Chile. Then there are five other *Lutra* species, all of which seem to follow the same sort of life style as our own otters. Some of these species occupy the same range as the giant otter, a fact that poses a rather interesting problem: all otters live primarily on fish, so how do the smaller otter species manage to live alongside their larger relative when zoological theory states that no two species with the same habitat requirements for food, sleeping areas etc can co-exist for any length of time? It was a problem that Liz was keen to answer.

In general, the giant otter conforms to the same anatomical blueprint as the rest of the Lutrinae, the sub-family to which all otters belong. The body is long, sinuous and unbelievably supple, and ends in a thick tapering tail by means of which the otter moves and steers through the water. The legs are short, and the five-toed, webbed feet bear sharp, curved claws which help to grip fish securely and make climbing slippery banks easier. The characteristic bullet-shaped head of the otter is a streamlined adaptation to its aquatic mode of life. The eyes and nose are positioned well to the top of the head so that the otter can still breathe and look around while the rest of its body remains sub-merged. The front of the face is covered with numerous groups of whiskers, especially useful when the otter must hunt in muddy water or at night. Even minor details are taken care of: both the otter's ears and nostrils close down the moment the head sinks beneath the water. As the naturalist, Bjorn Kurten, has said, 'the otter represents the most elegant solution to the problem of constructing an amphibious carnivore that Nature has so far effected'.

There are, however, several differences between the river wolf and the rest of the South American otters. Size apart, the giant otter, with its chocolate-brown pelt, rounded, dog-like head and thickly webbed feet, is easily distinguishable. The tail, too, is different, being flattened like a beaver's as an adaptation to its strongly aquatic existence.

While the anatomy of the giant otter was fairly well known, its behaviour was little understood. Victorian travellers had told of an eminently approachable animal, far less retiring than its European or North American cousins. The giant otter was said to earn its living during daylight hours, moving about in family groups of four, five or six and sometimes coalescing into 'super-groups' of up to thirty animals. Osbert's forty animals were either a record or the result of a little Caribbean exaggeration, a common complaint as we discovered later. The otter groups were reportedly quite fearless of man and inquisitive to the point of nosiness, approaching very close whenever a boat surprised them on the river and scrutinising the occupants. While this sounded perfect for the scientist intent on studying the giant otter, such behaviour was also ideal for any would-be collector, and it went some way to explain the large number of otter pelts in the museums and universities of America and Western Europe.

Like other otters, the river wolf's skin is covered by fur of surpassing beauty. The secret of the otter's pelt lies in its double-layered composition. There is an underlay of fine dense fur, protected by an outer covering of longer, stiffer guard hairs which are kept well oiled to repel water. So dense is the under-fur that even if the hairs are forcibly parted, the skin does not show through. As far as the otter is concerned, its fur is utterly functional; the dense under-fur holds a protective layer of air between the otter's skin and the surrounding water. But for Man, or rather for Woman, the appeal of an otter skin is purely aesthetic. Tanned, the pelt makes a wonderfully soft, warm hide, ideal for jackets and coats. Being the biggest pelt available, the skin of the giant otter is much sought-after by furriers who are willing to pay up to US $500, as much as for a jaguar skin, for a single perfect pelt. The demand – mainly from the affluent North American, European and Japanese markets – is so great that the river wolf is now under extreme hunting pressure throughout most of its range.

Despite recent legal protection in most habitat countries the situation is so appalling that in March 1977 and again in 1980, the Otter Specialist Group of the IUCN's 'Survival Service Commission' designated the giant otter as the most highly endangered otter species in the world, and gave it priority of funding. In Brazil, for example, the river wolf has been decimated by illegal hunting for the fur trade. The pelts are shipped illicitly from Brazil to Colombia and thence to the furriers of Paris, the skins of the giant otter ending their days draped across the wrinkled shoulders of some rich Western matron. And Brazil is not unique. In Peru, Ecuador, Paraguay and Venezuela, the story is the same: hunting, and to a lesser extent human population pressure, have all but destroyed this unique and beautiful species. Only in the extreme north-east of its range does the giant otter exist in anything like its former numbers, in the newly independent countries of Surinam and Guyana.

Thanks to a film on which I had worked for Survival, I knew a little about Guyana, enough to know that the land itself was incredibly photogenic. Countless rivers and creeks cut a tortuous, winding path through the rain forest that covers most of the country's 215,000 square kilometres. The name of the country derives from the Amerindian words 'Guaya Waina', meaning 'Land of Many Waters'. Guyana is not yet heavily populated by man, and more than ninety-five per cent of its estimated one million citizens live on a twenty-mile-wide coastal strip which includes the capital, Georgetown. The remaining 100,000 souls are scattered over the rest of the country, either on the cattle-ranching savannahs of the Rupununi or prospecting for gold and diamonds in the forests of the interior. As a result, animal life abounds and the almost virgin forests are peopled with hundreds of exotic creatures. Jaguar are so common that dogs are often taken by them from the outskirts of Georgetown. The other cat species are there in strength too, margay, ocelot and the secretive jaguarundi. There are otters, tapir, agouti, the mermaid-like manatee and seven species of monkey. There are innumerable birds of prey, including the ornate hawk eagle, harpy eagle, snail kite, laughing falcon and scores of smaller raptors. Snakes and various other 'nasties' abound: the Dendrobatids, or poison-arrow frogs, from which the Amerindians extract a lethal chemical to coat their arrows and blow-darts, anacondas, rattlesnakes, fers-de-lance, the equally poisonous coral snake and, most feared of all, the deadly bushmaster, whose 1½-inch fangs are the largest of any known snake. The list was almost endless, and very few of the animals had been properly filmed. And to cap it all, Guyana was a former British Colony and everyone spoke English!

Three months after our chance meeting with Osbert, our prospectus was complete and we were ready to try to sell the idea to one of the wildlife programmes. It seemed natural to ask Survival first, but to our astonishment, Colin Willock, the Survival producer, was not interested in the giant otter or our plans to film the beast. Our confidence considerably shaken, we showed our idea to the BBC Natural History Unit at Bristol. The difference was amazing. Richard Brock, producer of the much-acclaimed 'Life on Earth' series, introduced us to Peter Bale, Series Producer for 'Wildlife on One', the BBC premier animal half-hour. He was extremely enthusiastic and, within an afternoon, we had been offered much support and advice and, best of all, a contract! Two days later we received word that Liz had been accepted as a

PhD candidate at Cambridge, her subject 'The Ecology of *Ptero-nura brasiliensis*, the Giant Otter'. Once we were sure that the Guyanese government would allow us both to study and to film the beast, we took the plunge. Both resignations went in on the same day and with that there was no going back. We were on our way.

3   In the Land of Many Waters *(Keith)*

Ever since my camping days with the Scouts, my feelings about trips into the Great Outdoors have been divided into three definite phases. The initial planning phase, conducted in the comfort of one's own room, is a time of high enthusiasm and eagerness to be off. It's all armchair travelling, anticipation rather than realisation. During phase three, when I have actually arrived in the field, I tend to enjoy myself or, if the going is rough, to put my head down and curse through the trip as best I can. But in between these two times is a short, seven-day period – the Despondency Phase. It never fails; a week before the trip I decide that, really, I don't want to go. After all, why should I go? Don't I have everything I want at home? Aren't all my friends here? Why should I go risking my neck miles from anywhere, among strangers, when I could stay at home where (skinheads, terrorists and psychopaths apart) I would be free from danger? What am I doing? Waking and sleeping, these and similar depressing thoughts bounce around my brain. During earlier trips I used to be very worried by these unwelcome doubts and fears, but after a while I came to accept them as a necessary part of travelling, at least for me.

In this case there were other pre-occupations. Some time before leaving England, Liz had heard that an American naturalist, Nicole Duplaix, had recently returned from a preliminary study of the giant otter in Surinam, Guyana's eastern neighbour. Nicole knew her otters – she had spent some time as co-editor of the prestigious *International Zoo Yearbook*, and had studied many otter species in captivity. Liz wrote to her and discovered that Nicole was visiting London a few weeks before our departure, and would be happy to meet and discuss the findings of her trip. The talk proved illuminating, but what Liz heard was both useful and worrying. Nicole had confirmed many of the old travellers' tales about the river wolf: the animal was diurnal, and it did occasionally move about the river in large groups; the greatest number Nicole had seen together was sixteen. Normally, however, the giant otters seemed to live in family groups, a male, female and between one and three cubs. Occasionally, the cubs from the previous mating stayed on with mother and father and were thought to help in the raising of their younger brothers and sisters by catching food. The giant otter, said Nicole, was a loud and curious animal and this had helped her study enormously. She had also completed a preliminary survey of the giant otter throughout Surinam, and this had shown that the animal was still to be found in most of the main river systems in Surinam. Like ourselves, Nicole was very keen to see a similar survey made for Guyana; conservationists would then have data on the animal's distribution in the two countries that were thought to be the last remaining strongholds of the river wolf.

All these thoughts and worries were in my mind as I felt the aircraft bringing us to Guyana tilt round in a gentle arc to begin the final descent into Timehri Airport. Suddenly, instead of an azure canopy with small white clouds scudding lazily across the sky, I found I was looking down on a giant sea of green. To one side the green mantle was broken by a curving snake of grey-brown water – the famous Demerara River. The only signs of humanity were the airport landing strip, a few scattered buildings and, almost imperceptible in the gathering sunset, a ribbon of tarmac winding its way down the right-hand bank of the Demerara, more often than not lost among the blanket of verdure. For the first time I began to realise just how vast and untouched a land the Republic of Guyana was. To read that ninety-five per cent of the country was virgin forest was one thing, but to see the almost infinite scale of the vegetation from this 'god's-eye-view' several thousand feet in the air made one realise just how small an impression the hand of man had made upon this immense land. For as far as the eye could see the land rolled away in an unbroken vista of tall trees, their green tops like the flower of some gigantic broccoli. Mile upon mile of jungle, and it began not 500 yards from the airport terminal. I was excited and awe-struck; if this was the coast, what was the real interior like?

Only one area broke the unremitting monotony of green. A vast lake, looking like an enormous silver mirror in the dying rays of the sun, lay about a mile or two from the eastern bank of the Demerara River. This was one landmark I did know, thanks to the maps we had scrutinised so diligently in England. The 'mirror' was Russel Lake, 200 square miles of swamp and lake that had supplied Georgetown and the coastal villages with all of their water since the late 1800s. Emergent trees, standing thirty or forty feet above the canopy, obscured its shiny surface as our plane made the final approach into Timehri.

Half an hour later, along with a hundred other passengers, we were waiting in one long, untidy line at the only manned customs desk. As we got closer to the solitary official, one question, 'Drugs?', was repeated with mesmeric regularity. Guyana had no real drugs problems and, as the sign above the exit reminded everyone, the government intended to maintain this situation. A second sign, at floor level, informed the visitor of Guyana's list of banned items. Line upon line of close-packed print detailed the taboo articles: potatoes, onions, apples, sardines, corned beef, spam (in fact any type of tinned food was verboten), chocolate, dried fruits such as raisins or dates, jams, honey, biscuits of any description. In short, any merchandise that gave comfort and pleasure to westernised man seemed to be included on the list.

Luckily, most of our preserved food was freeze-dried, as we had been told by the Guyana High Commission in London of the prohibition on bringing tinned foods into the country. It seemed a harsh law, but we were very sympathetic to the government's motives. Like many newly independent states, Guyana wanted to decrease its dependence on foreign-made goods, to become much more self-sufficient. Until independence, British-style food had been freely imported into the country and the Guyanese had developed a liking for Anglicised dishes, made from just those types of food that could not be produced within the country: potatoes, nuts, corned beef, grapes, even chestnuts! The Guyanese government was trying to re-educate the population's eating habits and so save valuable foreign exchange. But unlike Britain, there were no subtle propaganda battles to alter public attitudes. The guilty items were simply banned, and smugglers jailed.

We had been warned about tinned foods but not about biscuits. I suddenly realised that I had a problem. Deep in the recesses of my largest piece of luggage, carefully packed to avoid damage, were six packets of what I now knew – in Guyanese terms – was contraband: MacVities Digestives! I should declare these wicked capitalist luxuries, but the thought of going six months without a single wholemeal biscuit was too much. After all, I rationalised, I wasn't Guyanese, and provided I didn't corrupt the population by passing round the digestives at secret midnight biscuit-parties I was surely justified in keeping them. The role of the smuggler had always appealed to me – sneaking through customs with £100 000 worth of gold or diamonds had always seemed a romantic way to make a living. Now I was embarking on my life of crime with contraband biscuits! Fortunately, the single customs

21

man was tiring by the time he reached us towards the end of the line, and he waved us through with scarcely more than a glance at our bulging cases. The illegal digestives were safe.

Outside, the short tropical twilight had all but given way to night. All around towered tall trees; the jungle proper seemed to begin just beyond the airfield perimeter, and the night was alive with the bell-like tinkle of tree frogs and the harsh shrilling of cicada. A palm-lined road snaked away to the left and was lost among the dark shadows of the forest. It was the Hutchinson Highway, the only road, and it led to Georgetown some thirty miles away. Small, dark knots of people were lazing their way along the road, stopping occasionally to let pass what seemed to be miniature sandstorms. As they approached, these mini-hurricanes resolved themselves into vehicles, mainly lorries and Morris Oxfords of antique vintage. We were still taking all of this in when one of these dust clouds detached itself from the main convoy and a huge American gas-guzzler, a '57 Cadillac, powered towards us at 60 mph, braking to a halt a few feet away. A grinning negro face appeared around the grime-obscured windscreen. 'Taxi?' said the face.

'How much to Georgetown?' Liz countered. We wanted to get to the capital as soon as possible, but we weren't so green that we didn't ask prices first.

'I say later,' answered our grinning guide.

'No way, man,' Liz replied, lapsing into a broad Barbadian accent. 'I no tourist to de parts down hey. I'se Bajan.'

'Fuh true?' the cabman asked, taking in her blonde hair and pale skin. He let out a series of tuts, as if pondering this unexpected information, then looked up. 'OK,' he said briskly. 'To Georgetown will cost you eighty dollars Guyanese.'

This sounded reasonable to me, and I had already stepped forward with the baggage when Liz coolly replied, 'Forty dollar.'

I thought we had lost our cab, but the driver was not in the least put out. 'You *is* Bajan!' he exclaimed, slapping his thigh in delight. 'Them jes' 'bout de meanest people I know.' He opened the door wide. 'OK, little lady, fifty dollar.'

Liz inclined her head in agreement, then turned questioningly to me. 'What fuh yuh waitin', Mr Tourist?' she asked. 'Load de baggage!'

The female giant otter looked absolutely beautiful. She was so close we could see her haired nose (unique among the world's otters) and the long whiskers that covered her face, looking like bunches of silver wire close to her bullet-shaped head. At this

22

distance we could also see clearly the smooth softness of her dense, chocolate-brown fur. She was dozing, and the only thing that marred our mounting excitement at seeing a live giant otter for the very first time was the knowledge that she was a captive, locked in a small cage in Georgetown Zoo.

Her name, the keepers told us, was Squeaky. She had been brought to the zoo eight years before by an animal exporter who had raised her from a cub until Squeaky's natural otter exuberance and extra-large size had made her unmanageable. We had been at Squeaky's cage for hours, trying to absorb the essential aspects of the giant otter so that we could more easily recognise her relatives in the field. The shape of the head – probably all that we would see of a swimming giant otter – was especially important. We wanted to be able to recognise the head from any conceivable angle: how it was held in the water when swimming, the way it moved before a dive or when it was angry, afraid, or just plain inquisitive. We wanted to watch the otter moving about on land and to know the animal's calls in case it was at any time hidden from view. Any posture or mannerism that might allow us to differentiate the giant otter from other aquatic animals was valuable, especially as South America has more than its fair share of swimmers.

Because of the large numbers of creeks, rivers and lakes in Guyana, and because many of the country's forests are prone to serious flooding during the rainy season, a large number of animals have learned to use swimming as just one of several means of transport. Knowing this, we had spent time familiarising ourselves not only with Squeaky the otter but with several other animal species that might cause confusion if seen at a distance in the water. The bush cow or Brazilian tapir (*Tapirus terrestris*) is always found next to creeks and rivers, often taking to the water if pursued by hunters or animal predators like the jaguar. When swimming, only its long mobile snout and perhaps the top of its head are seen, and to the inexperienced eye it might easily be identified as an otter. Such confusion is even more likely with the jaguarundi (*Felis yagouaroundi*), a member of the cat family that looks less like a cat than any of the world's felids. In certain areas of South America it is even known as the otter-cat. Fortunately for us (though not for the otter-cat), the jaguarundi is both nocturnal and uncommon in Guyana, so cases of mistaken identity would, we hoped, be rare. The same could not be said for the caiman, a relative of the African crocodile and the American alligator which is still common in the creeks. Although the Guyanese caiman can reach lengths of more than six feet, such specimens are now rarely caught. Man has killed off all the bigger individuals in his search for 'crocodile' skin. In some parts of Guyana, we were told, the local fishermen had exterminated even the smaller caiman because they occasionally steal fish from seine nets under cover of darkness. But, instead of taking more fish in their nets, the fishermen found that their catches gradually decreased after the extermination of the caiman, until eventually they were bringing home hardly any fish. The reason was far from simple, but when finally unravelled it served as an object lesson against meddling with Nature's precarious balance. The caiman droppings fertilised the creek waters, releasing nutrients that microscopic plants, known as phytoplankton, used to feed upon and increase in numbers. The now numerous phytoplankton became food for zooplankton, tiny, free-floating animals which were in turn consumed by the fishes, upon which the fishermen lived. No caiman faeces meant no phyto- or zooplankton, which meant the fish starved to death and were simply not there to catch.

But of all Guyana's animals, by far the most serious problem as far as identification was concerned was the Guyana otter, *Lutra enudris*, a smaller otter species that by all accounts lived cheek by jowl with the giant otter in the rivers and creeks. There was a captive *Lutra enudris* at the zoo and we took pains to study this species' features almost as well as we had studied the giant otter's. Close to, it was easy to distinguish the two species – apart from other differences, at three to three and a half feet long, a full-grown *Lutra* was only half the size of the river wolf. But I was less sanguine about our ability in the field. Would I feel so sure when I encountered a black-grey silhouette a hundred yards downriver on a misty morning in the interior? Liz must have been thinking along the same lines, for she turned from the captive otter, muttering to herself, '*Lutra* is going to be one hell of a handful.' Perhaps she was psychic, I don't know, but it wasn't until much later that we realised the truth of Liz's prophecy and came to know just how much of a handful a Guyana otter could be.

This was our fourth day in Guyana. We had spent the first day negotiating with the zoo to build an enclosure in which we could more easily film close-up and underwater shots. The rest of the time had been taken up with frustrating journeys between the police Immigration Department, the Ministry of Home Affairs and the Interior Ministry. Although we had already obtained permission for our trip while in the UK and had letters of introduction to prove it, we discovered that the Airport's immigration stamp in our passports entitled us to stay in Guyana for four days

24

only. We rushed off to the Ministry of Home Affairs and were only slightly surprised to discover that they knew very little of our proposed expedition. We were obliged to run through the whole rigmarole again, visiting ministers, under-ministers and their secretaries and putting details of our intentions in writing and in triplicate. The letters disappeared into the bureaucratic machine and the wheels ground round slowly. It was a terribly frustrating time. The only pleasant interlude was our obligatory registration at the British High Commission. After the formalities had been completed, we were ushered into the presence of the High Commissioner himself, Mr Philip Mallet. Over afternoon tea we discovered that the High Commissioner, a slim, elegant man in his fifties, was very knowledgeable about Guyanese wildlife. Or rather, one aspect of it. He was quite obviously a bird-man, and while he told us of many areas where interesting bird species could be easily photographed, he seemed to show a complete lack of interest in the giant otter. In the High Commissioner's mind, birds were the only wildlife worth watching, and the half-hour we spent with him was full of interesting snippets of information that we stored away for future use.

Fortunately for our sanity, the bureaucratic waltz ended during the afternoon of our third day; the presence and purpose of our expedition was fully endorsed, albeit for the second time, by the Guyanese Government. On reflection, I wondered if the British Civil Service would have been so efficient!

Clearing this final official hurdle took such a weight off our minds that we felt as if we were on holiday instead of being right at the beginning of an arduous twelve months. For the first time since we arrived we had time to notice how beautiful a city Georgetown was, the wide, sunny streets lined with three- and four-storeyed wooden buildings executed in Dutch colonial style with their low-eaved roofs, louvred windows, low, shady verandahs and gingerbread gables. We strolled down the Avenue of the Republic, past the Law Courts on one side and shrill mobs of homeward-bound commuters on the other, crammed nine to a car for their daily ride to the outlying villages. We wandered through the Guyana Museum, looking at the wide variety of Amerindian artefacts displayed there, reed haversacks, hammocks, fish-traps and blow-guns. From there we visited the world's largest wooden building, the massive white bulk of Georgetown Cathedral. Built entirely of local timber and touching 150 feet at the pinnacle of its timber spire, the vaulted interior was cool and shady and beautifully decorated with carving and filigree iron-work, cunningly wrought. Then, our minds refreshed, we sauntered back to our hotel in the cool of the evening. It was then that we were robbed.

We ought to have been more careful. It was common knowledge that the city was infested with 'choke-and-rob' bandits whose main prey was foreign visitors. Choke-and-rob men come up on their victim quickly and silently from behind. The hit squad is divided into a rob-man and one or two choke-men. Choke-man number one immobilises the target's hands, while choke-man number two pinions his legs. The rob-man swiftly searches his pockets, helping himself to anything valuable: wallets, rings, watches, necklaces, medallions, even the victim's sunglasses. When the rob-man has finished, the choke-man, with practised ease, throws the victim to the ground and they make off. Most heists take place close to Tiger Bay, a rabbit warren of narrow alleys and ramshackle dwellings that stand, incongruously, right next to the most modern hotels in the country. We had heard harrowing tales of fingers cut off by the choke-and-rob men to obtain gold rings, of earrings snatched from the ears, ripping the victim's lobe in two, and many other horror stories.

It all began so innocently. Our route took us down Main Street, an impressive thoroughfare flanked by large, colonial-style houses. The road itself is dual carriageway, with a central grass walkway, a filled-in canal from the early Dutch days. Our hotel stood on the left of the road and, without thinking, we crossed over to that side some 150 yards before the hotel entrance. Suddenly, as if they had grown out of the ground, there were three tall negroes about fifteen yards behind and walking so fast that they must overtake us before we reached the hotel. It was a typical choke-and-rob set-up, and to avoid it we stepped into the mouth of a short alleyway and made as if to talk, hoping that they would pass us by. They didn't. Instead, the trio ranged themselves across the entrance. We retreated further down the alley, which turned at right angles about fifteen yards from Main Street. From there we would have a straight run for the hotel's rear entrance which, we knew, lay opposite the alley. But a shock awaited us: at the bottom of the alleyway was a high gate. We turned to find the three men slowly advancing on us, completely blocking our exit.

While healthily scared, I did not feel our position was completely hopeless. Four years training in karate had ended with a back injury, but not before I had earned a brown belt and the ability to kick and punch faster than most untrained men. In addition, I was carrying in my pouch a nun-chaku, two foot-long staves of hardwood joined together by a short chain. It's a simple weapon, but it can deliver a vicious and often deadly blow. Hidden in her umbrella, Liz had a long ice-pick with which, by dint of constant practice, she could skewer two-inch potatoes thrown in the air. But even if we won this street battle, I was still mortally afraid of the outcome. For two foreigners to use offensive weapons to hurt seriously three Guyanese nationals would put us in a terrifying predicament. The best that could happen would be expulsion from the country, the worst . . . well, the worst was best not thought of. Ordinarily, I would have happily handed over my ready cash, watch, even my shoes. But today, because we had visited the immigration office, I was carrying in my pouch our passports, travellers cheques and nearly US $800. If we lost all that we might as well say goodbye to the whole trip. It would be the end of our dream. We couldn't let that happen. We had to fight.

At that moment another figure appeared at the top of the alley, silhouetted against the light from the street lamp. My heart sank. Now it was four to two. The fourth shadow spoke.

'Hey, man,' it said, 'what you three boys playin', nuh?'

Our three potential muggers froze. They looked uncertainly amongst themselves, and I felt my spirits rise with their confusion. Had we found an ally?

'We ain't doin' nuttin',' the tallest mugger said, looking round at his comrades who nodded their agreement. 'We jes' wanted a light for de cigarettes.'

'The gentleman don't smoke,' said the fourth shadow. In the confines of the alley the words rang eerily. He moved to one side of the alley. 'So why don't you boys just run along?'

To our astonishment, that was just what they did, shuffling off meekly past our new-found friend who stood with his back to the wall and his right hand tucked in under his shirt. We walked back into the bright lights of Main Street, though I still kept a tight grip on my nun-chaku; was our 'friend' merely chasing off the competition?

We were soon reassured. Our rescuer, who introduced himself as Martin van Hoof, accepted our offer of a drink at the hotel bar. He was a dark-faced hawk of a man, with a flashing smile marred only by the fact that his front two teeth were missing. Though a man of medium height, Martin was extremely heavily built, and all of his bulk was muscle. Noticing the nun-chaku, he asked if I was interested in budo, the Japanese martial arts. I told him of my experience and, over a large glass of ice-cold beer, Martin said that he was a budoka too, a black belt in judo.

'Was that why those three men backed off?' Liz asked. 'Because they know you're a black belt?'

27

'Maybe,' Martin replied, looking round him from under hooded lids. 'But I think perhaps they knew that Little Brother was watching too.' He lifted up the front of his shirt, and in the waistband was the squat, ugly shape of Little Brother – a .357 Magnum revolver. No wonder the muggers had fled so meekly!

'Is that legal?' I asked uneasily.

Martin eased my worries considerably by saying that he was one of a handful of civilians authorised to carry a handgun. After regaling us with tales of his anti-mugging career, Martin asked us the purpose of our visit. He was intrigued by our answer, saying that he too was very interested in wildlife, more especially in reptiles, which he exported to pet dealers and venom laboratories in the United States. Suddenly, the few facts I knew about Martin – judo, revolver, reptile exporting – fell into place and I knew the identity of our rescuer. He was someone we had been trying to contact almost since we arrived. Martin was the 'Snake Man'.

One of our major problems when we had first arrived was to decide which part of the country we would visit first. It had to have lots of giant otter in it, as we planned to use this area to learn more about the animal. First we would look at the river wolf in captivity, then in the wild in an area of relatively high population. Here we could study the animal's life style in detail. Then we could move into areas in which the giant otter's population size was not known and, using the knowledge we had gained earlier, estimate just how many giant otter survived in these 'unknown' areas. The zoo would give us our captive study, but where in Guyana was the best place to see wild giant otters? When we tried to find out we were told time and again to 'talk t' de Snake Man'. He travelled extensively in the interior, but more than that, the Snake Man had reptile collectors scattered throughout the length and breadth of Guyana, who regularly sent in supplies of snakes, lizards and frogs. They were as efficient a spy network as one could find in Guyana and, we were assured, the Snake Man would soon be able to tell us where the giant otter was to be found in the greatest abundance.

And now we had found the Snake Man, or rather he had found us. And he did have a number of contacts that might be of help. Offhand, Martin thought that the best area for giant otter was the Upper Mazaruni Basin, an isolated area which bordered Mount Roraima, the massive flat-topped plateau famous as the inspiration for Arthur Conan Doyle's novel *The Lost World*. Because of the danger and hardships involved in getting there, hunters were scarce and the giant otter populations healthy. If he were betting money, that is where he would head first. But, as Martin was quick to point out, the very points that made the Upper Mazaruni such a good otter habitat were also disadvantageous to a long-term study and filming trip such as ours. The logistical problems in staying for an extended period in the Upper Mazaruni were formidable. Everything would have to be flown in at enormous expense, and any exposed film would have to leave for Georgetown by the same route before the humid jungle air allowed fungus to feed off the precious emulsion.

Martin swallowed the last of his drink, looked at his watch and walked off into the night, his right hand resting easily on the waistband of his trousers, next to the Magnum. We were now firm friends with the Snake Man, and that fact alone conferred on us a great degree of security, at least in Georgetown. Anyone who rolled us would have Martin and his Magnum to contend with. What was more, the next day Martin phoned around his friends and, with a few dissenting voices, the consensus was that the Upper Mazaruni Basin did hold the highest number of giant otter. Other than that, we could try Spectacle Lake on the Rupununi, which was also said to possess a healthy population of the river wolf. Both areas were equally remote so, in the absence of any better information and despite the problems of travel, we decided that the Upper Mazaruni was as good a place as any to make our debut as giant otter watchers. Having made our travel decision, and feeling much safer in Georgetown's streets, we headed for the zoo with renewed confidence.

That confidence, and the six hours we spent watching Squeaky the giant otter, greatly restored our enthusiasm for the trip. We were still studying the small Guyana otter when a voice hailed us from across the zoo. Turning, we saw the thin figure of the British High Commissioner, resplendent in tropical shirt and shorts, standing by the tapir enclosure and waving us over.

'Still looking at otters, I see,' he said as we approached. 'But that's not the giant form y'know.' He pointed over towards Squeaky's cage. 'The big chappie's up that way.'

'Yes,' Liz replied, 'we've been watching it for hours. It really is enormous, isn't it?'

'Oh, indubitably,' the High Commissioner answered airily. 'But that's not the biggest I've seen. Why, only a fortnight ago, and not thirty-five miles from here, we saw five of the beasts, at Lama actually, and one of them . . .'

'You saw five?' Liz broke in excitedly. 'Whereabouts? Where's this Lama place? Were the five a family group or were they . . .?'

This time it was the High Commissioner's turn to interrupt. 'My dear girl, patience! All in good time. Lama is a rather primitive rest-house owned by the EDWC on Russel Lake.'

'That's the East Demerara Water Conservancy?' I asked. 'And Russel Lake's the huge stretch of water you can see when you're coming in to land at Timehri airport?'

'Quite so,' replied the imperturbable diplomat, icy calm in the face of our barrage of enthusiastic questions. 'As I was saying, Lama rest-house lies on Russel Lake, and it's sometimes possible to arrange to spend time there to, as it were, recharge one's batteries. No-one's allowed inside the conservancy and the place is so calm and quieting for one's nerves. You do follow me?'

We nodded our agreement and the High Commissioner drew in a deep breath before continuing. 'Excellent,' he said. 'Well, there are canoes one can borrow to paddle down the creeks and, every time my wife and I go there – and we've been to Lama this year already – we see these large giant otters. On our last trip we saw five, two adults and three young, and one of the blighters, a terribly big chap, came so close to the canoe that Mary, my wife y'know, she had to chase it off with a paddle!' A thought seemed to strike him and he grew quiet. After a moment he continued, 'Y'know, you really ought to go up there if you're so interested in the damn' beasts. There's lot of birds too: jacana, osprey, snail kites, black-crowned night herons, even the odd hawk eagle. And on our last trip we saw the nest of an absolutely fabulous bird, a . . .'

I could see that the High Commissioner was slipping back to his favourite subject so, at the risk of seeming rude, I interrupted again. 'But are you certain these otters were giant otters?'

'. . . a lesser spotted . . . What? Oh quite sure,' the High Commissioner replied, looking almost offended. 'The one that charged our canoe was this size.' He held out his hands like a fisherman describing his catch. 'No, I'm wrong,' he insisted, 'it was even larger', and he stepped backwards and extended his arms even wider, emphasising the animal's huge size.

This was unfortunate, as the High Commissioner was already standing on the edge of the tapir mucking-out pit. As we watched in horror, he teetered for a moment on the brink, his arms still spread and a look of mixed surprise and indignation on his face, before collapsing backwards into a mixture of mud and manure. It was a most embarrassing moment – what do you do when Her Britannic Majesty's Representative to the Co-operative Republic of Guyana falls into a muck-pile? Fortunately, Mr Mallet's diplomatic training once again saved the situation. Springing up and wiping the evil-smelling ordure from his soaking shorts, he said briefly that the accident was 'Damned annoying' before carrying on his conversation as if nothing untoward had occurred. He was adamant that we try Lama and the EDWC first before venturing so far into the interior as the Upper Mazaruni. A cloud of flies, attracted by the fruity odour, began to gather over the High Commissioner's head in a black swarm but he bravely ignored them. He would get clearance for our visit from the EDWC authorities. Would we like to go?

In a situation like that, how could we refuse?

4 Lama

(Keith)

In the early days of Georgetown, before the 1820s, the settlement obtained its water supply from a few inadequate wells. It took an indefatigable Scot, William Russel, to conceive the idea of damming the rivers around Georgetown to create a giant reservoir sufficient for the town's needs. Although William Russel did not live to see the completion of his life's work, the lake that finally rose behind the mud walls of his dams showed that his original idea had been faultless. When full, this immense reservoir, covering more than 150 square miles, was able to defy the worst droughts that Nature could send against the Guianas. In honour of its creator it was named Russel Lake.

For the wildlife of Guyana, Russel Lake was something of a mixed blessing. Long stretches of riverine forest and seasonal swamp were drowned as the waters backed up, and with the habitat the wildlife inhabitants perished. Innumerable bird species, howler monkeys, capuchins, ocelots, tapirs, carnivores and herbivores of every description, all went down before the inexorable growth of the lake. And yet for other species it was as if a promised land had suddenly opened to them. Lagoon, lake and swamp-living species converged on the huge stretch of water, full of fish and offering ample room for expansion. Herons, spur-wing, harriers, fish eagles, capybara, caiman and giant otter were all drawn to Russel Lake. It was a magnet for water-bound animals.

We discovered this almost as soon as our speedboat had left its mooring at Land of Canaan, a small village next to the Georgetown–Timehri road. Because it was totally isolated from roads or even tracks, the easiest way to reach Lama rest-house was by crossing the whole of Russel Lake, some twenty-seven miles, in a speedboat. With its smell of petrol engines, its jetties, outbuildings and artificially straightened canals, the hand of Man lay heavily upon the Land of Canaan. But all this changed as we left the straight, half-mile stretch of canal that took us into Russel Lake proper. It was like entering a different world. Stands of

32

moca-moca lined the banks, their long, two-inch-thick stems ending in foot-sized arrowhead leaves and growing in such thick clusters that it was impossible to see more than two feet on either side of the creek. Elsewhere, where the water was too deep for the moca-moca, extensive reed beds swayed gently in the breeze as we sped past. Here and there tall thin Ite palms (*Mauritia flexuosa*) stood like lonely sentinels among the fastness of the lake. These palms are specially adapted to swamp conditions and will find a foothold in the smallest patch of shallow water. On the horizon groups of Ites, looking for all the world like miniature feather-dusters at this distance, marked out such shallow-water areas like desert islands on a flat sea of green.

After almost ninety minutes travel in the speedboat (not counting eight stoppages to remove weed and water lilies from the plant-choked engine), the flat, Ite-dotted horizon became green and dark with the typical crenellated silhouette of riverine forest. We watched this point closely, knowing that we must be close to journey's end by now, and eventually Liz pointed out a small building on the horizon to the right of our boat. Fifteen minutes later we were close enough to pick out details of a building and were pleased to see that it was a two-storey structure and not simply the miserable little hut we had been led to believe while in Georgetown.

When we reached the covered 'dock' that fronted Lama, a long, thin sapling of a man with a face like a machete appeared as if by magic from behind one of the wooden support posts. This, we were to find later, was Katina, the watchman of Lama and a fair contender for the title of Dourest Man in Guyana. Katina tied the boat to a rusty iron hook set in an upright and we stepped ashore.

The introductions were over very quickly: Katina the watch-man and Pappy, the pint-sized cook, were, with the speedboat-driver, whose name was Stanley, the only EDWC employees regularly stationed at Lama. The three men, all of East Indian extraction, helped us to carry our cases into the rest-house. I went first, through the gate of the ground-floor fence – and stopped dead. Precisely in the middle of the rough wooden floor, extending from one end to the other, under tables and chairs and neatly bisecting the room, was a six-inch wide column of *Eciton hamatum*, the infamous red-brown army ant. The column seethed with activity, some ants moving from left to right and the remainder bustling in the opposite direction. On either side of the red-brown mass was a thin line of what appeared to be miniature yellow bulls. When one looked closer, it became obvious that only their heads were yellow, but what heads! They were completely

33

out of proportion to the rest of the body, at least three times as large as those ants in the central ant column, and sprouting from the front of each overgrown head was an enormous pair of wicked-looking curved jaws – the horns of the miniature bull. These were the soldiers, one of three castes in army ant society which, apart from the Queen, consists of two different types of worker (the major and minor castes) and the soldiers. It was the workers that made up the central file of the main column, with the soldier caste taking up protective positions on the flanks.

To us newcomers at Lama, the watchman, driver and cook seemed to be taking this invasion rather quietly. In Ghana we had seen large troops of ants dispersed with bucketfuls of soapy water, and we asked why the same defence was not used here. We were astonished when the cook, whom everyone called Pappy though he was the youngest of the trio, said that he wanted them in the house, that they would not stay long and were very useful. Seeing our surprise, he led us out of the rest-house and round to the back of the building. Here was the head of the ant column, the vanguard spread out in an elliptical shape some five feet across and just beginning to forage at the front of an old wooden storehouse. Beckoning with his finger, Pappy took us round to the back of the storehouse and asked us to wait for a while and to keep our eyes on the storehouse. After a few minutes, a tropical cockroach, almost never seen in daylight, squeezed from between a crack in the storehouse and rapidly took wing. Another followed, and then a six-inch gecko, another night feeder, scampered in panic from the storehouse. These early escapers were followed by others, and in a matter of minutes the trickle of fleeing creatures had become a flood. Spiders, lizards, cockroaches, silverfish, beetles, a small snake, all streamed in mortal terror from the storehouse and either dropped to the ground and fled into the undergrowth or spread their wings and took to the safety of the air. Even the resident bats flapped out in alarm. I now saw the reason for Pappy's pleasure at the ants' coming: they were master pest-operatives, destroying all the unwelcome hangers-on, the uninvited guests that infested the average tropical house and made life miserable.

Fortunately, our bedroom – a bare room with two ancient iron bedsteads – had already been processed by the army ants, and once we had moved in our baggage, we were so tired that we turned in for the night. Next morning the otter hunt would begin in earnest

We rose just before dawn and took our tea onto the verandah

outside our room to watch the sun come up. The first streamers of light were just beginning to colour the eastern sky above the forest, but the creek and Russel Lake were still blanketed in darkness. Moths clustered everywhere around the bare, generator-fed lights of Lama, everything from tiny, centimetre-long *microlepidoptera* to gigantic *morpho* butterflies with wing-spans of eight or nine inches. A praying mantis swayed uncertainly under the bulb and two cockroaches, the first to re-colonise the rest-house after the army ants' onslaught, scurried in confusion over the wooden walls, seeking a crevice in which to pass the day. The sandflies – small, midge-like insects – were biting terribly and mosquitoes hovered everywhere, but most of the night creatures were still active at this time, especially the bats, and our discomfort was amply rewarded by the glimpse of jungle night-life our vigil offered.

As the darkness retreated, the night-shift workers began to clock off. The moths flew to their hiding-places, the square-winged silhouette of a night hawk flitted by, and bat after bat flapped its way up to a small hole in the galvanised roof above our heads. Each time one entered, the air would be full of chitters and angry squeaks as the bats fought among themselves for the best roosting-places in which to spend the bright hours of daylight.

Then suddenly everything stopped. It was uncanny, a small gap in time, a complete trough of activity between the night- and the day-livers. The timing was eerie: within one minute of the last bat entering the hole above our heads we saw the first heron of the day winging its way across the still-dark waters of Russel Lake. More followed, then two spurwings, their long legs and outsize feet trailing behind them, landed on the nearby lily pads, calling softly to one another. The egret flocks retraced their airy path of the day before, and a new day had begun.

After breakfast we tried our luck on Russel Lake in one of the rest-house's boats, a long, flat-bottomed canoe that did not take kindly to being propelled by two landlubbers having their first taste of the ungainly, spear-shaped Amerindian paddles. In one of the boxes we had brought with us was an outboard motor, but we planned to use that only for travel to more distant parts of the study site. For actually finding the giant otter we would use paddle-power. Over the next three days we gradually mastered the canoes and scoured the area immediately around Lama, trying both Russel Lake and the natural creeks to the east. But it was not until the fourth day that the fates proved kind and we at last came face to face with the group of wild giant otter and those infamous killer bees.

That first meeting with Lama Creek's four resident otters convinced us that the area was indeed a good place to become acquainted with the habits of the giants. Now we had to get to work. Our job was to act as otter sleuths, collecting evidence on all aspects of otter behaviour, sifting it and trying to come up with some illuminating facts on the giant otter's life-style.

Our first task, like that of all good detectives, was to identify each of our subjects by sight and to learn their routine and pattern of movement throughout a typical day. Over a long period of time we slowly attempted to become more familiar with the otter group we had first encountered. We studied their tracks in the wet mud of the creekside and learned that certain sorts of ripple on the surface of the water indicated that a giant otter, and not a tapir or anaconda, had just left the water for the land. But what we most wanted was direct observation. It was no easy task; at first, almost as soon as they were seen and felt threatened, our family of four made off, either losing themselves in the dense bankside vegetation or simply submerging with a soft 'plop!' beneath the opaque waters of the creeks and swimming off heaven knew where. As time went on, the otters slowly began to realise that although we might be something of a nuisance, we meant them no harm. They gradually became more and more accustomed to our presence and by the end of the fourth week we had had enough close encounters to have identified and named three of the group.

'Spotted Dick' was the adult male, named for the distinctive mottled appearance of his cream throat-patch. He was the largest animal in the group, with a thick neck and large head, features we later discovered to be characteristic of male giant otters. Spotted Dick was without doubt the defender of the group; he was invariably the first to call attention to intruders and to investigate the threat they posed. 'Black Throat' was his mate, slimmer than the male and perhaps slightly shorter in stature. Her sobriquet derived from the surprising absence of any cream throat-patch, a feature which as far as we knew had never been described in the giant otter. Black Throat was capable of being quite as aggressive as her consort, though she usually let her spouse make the running during preliminary encounters with our boat.

The two young cubs were much more difficult both to identify individually and also to sex, and it was not until the fourth week of the study that we were able to make even an educated guess as to their gender. The female cub's throat-markings were a single ring of cream-coloured fur just below the neck: we named her 'Ringo'. The male cub's markings were a mystery, and he earned

36

the name 'Anonymous', being so shy and retiring that we could not identify him with certainty. Often, we would see both parents and Ringo swimming along in front of our boat with nary a sign of Anonymous until he broke cover from behind the moca-moca when he decided that the rest of the family had put sufficient distance between themselves and our canoe.

On one occasion, about two in the afternoon, we watched our family group heading for the river bank, and heard them (we thought) moving ashore to rest. Sure enough, a few minutes later, we caught the sound of chirping and chittering in the bush alongside the river, just opposite where we had moored the canoe. Gradually the noise diminished to silence, and we imagined the otters curled up on the bankside leaf-litter, slowly drifting off to sleep. We decided to wait them out, confident of re-establishing contact and getting more data when they finally ended their siesta and re-entered the water. We were proud, too, of adding new vocalisations to the list of otter calls we already knew – those strange noises just had to be comfort-noises given when the otters were grouped together on land.

Four hours later there was still no sign of the otters. With cramped, aching limbs, dehydrated bodies and the first attacks from the creek's mosquitoes, we admitted defeat and turned the canoe for home. The otters' cat-nap had become a fully fledged coma and we saw no chance of them returning to the water in the coming dusk. Still, it was all grist to our scientific mill. No-one before had recorded this strange phenomenon of giant otters retiring so early to bed. It was one more piece in the jigsaw puzzle of their behaviour.

We were almost back at the rest-house when we heard it again, the otter's quiet sleeping sounds. But something was wrong: it was coming from a point in the trees some sixty feet above our heads! Visions of tree-climbing otters evaporated and our faces became distinctly red when we made out capuchin monkeys in the trees, chittering quietly to one another as they took a late supper from a fruiting fig tree. Our 'sleeping otters' had been phantoms! We had been listening to the receding sounds of a capuchin troop, the noise fading away to nothing as they moved further and further upriver. Where the real otters had gone was anyone's guess; they had probably swum off underwater within the first five minutes of our watch, leaving us to peer for four long hours at an empty stretch of jungle! Although it was embarrassing at the time, we did learn an important lesson from this fiasco: to study the giant otter effectively we would have to learn everything we could about the rest of the forest's inhabitants.

As it turned out, the capuchin monkeys were one of the easiest of the other animals to study, and our third meeting with this species proved particularly illuminating. We were some five miles down Lama Creek, resting after a morning's hard paddling and having had neither sight nor sound of the giant otter. Quite suddenly, from the trees downriver there erupted the most almighty commotion. We cast off from our resting place and drifted slowly towards the sound on the ebbing tide of the river. As we got closer we could see that the source of this cacophony was two troops of capuchin monkeys who lived on opposite sides of the river. For some reason, the leading male of each group had taken it into his head to face up to his counterpart across the river. So they had each ventured out onto an overhanging branch and were vigorously bouncing up and down on these boughs, each screaming angrily at its opponent. They were the picture of venom and fury, their erect fur making them seem twice the size of their followers, their brows knitted with anger and mouths open to reveal long, pointed canine teeth. One would sometimes retreat along the branch into the obscurity of the foliage, only to rush forward once more with renewed vigour to begin branch-bouncing afresh.

I don't know whether these simian war-lords realised it, but to us their posturings were a very obvious and elaborate sham. With forty feet of river between the two antagonists there was simply no way they could ever come to grips with each other; and safe in this knowledge, each could vent his spleen on the other, showing his respective troop just how brave and strong he really was. I'm not sure how successful these bully-boys were; from where we were sitting, the other members of the troop seemed utterly bored by the pugilistic proceedings. The leader of the 'right-wing' troop, whom we had seen the previous day and knew as 'Big Boy' on account of his muscular, beefy appearance, seemed to be the more truly aggressive, for he occasionally turned his attention from his monkey-foe and treated us to a mini-display while we sat quietly in our boat. Then it would be our turn to face his open-mouth stare and the impressive display of branch-bouncing.

Eventually, honour satisfied, the less belligerent male backed slowly down his display branch, still arrogantly facing Big Boy, and gradually melted away into the forest with his troop. But Big Boy seemed to feel cheated; he had certainly not exhausted his bellicose tendencies and he was loath to give up so easily. We were the only opponents left him and so, for the next ten minutes, we were treated to a spectacular display of bouncing, climbing, staring, branch-shaking and general name-calling. As we did not

38

respond to his threats, Big Boy became increasingly bolder, eventually running along a branch with such aggressive abandon that he lost his footing and fell a good six feet towards the river! And that would have been the end of Big Boy, a snack for a caiman, had he not been blessed with a prehensile tail that snaked around a moca-moca stem as he fell, leaving him dangling like an oversize fig only inches above the water. But like any good actor, Big Boy was quick to recover and, bobbing about on the moca-moca stalks, he managed to give a good impression of having planned the whole thing.

Big boy eventually led his troop off, chittering triumphantly at our cowardice. He must have felt wonderfully powerful, threatening those two huge creatures, so much bigger than himself, and they so cowed by his magnificence that they dared not try even a little boat-bouncing! We slipped our mooring – and paddled straight into that piece of good fortune that all field-workers pray for.

We heard them before we saw them, a throaty bark followed by three or four chortling grunts. Spotted Dick and Company! And headed towards us by the sound of them. As silently as possible, we turned the canoe and made for a small creek that ran off at right angles from the main stream, overhung with branches of the aruba tree. Safe in this shadowy cave of vegetation, we turned expectantly to the far end of the canoe, nearest to the main creek.

We were only just in time, for no sooner had we got ourselves settled than around the corner sailed our family of four. Spotted Dick, as befitted his position of guardian of the family, was firmly in the lead, with Black Throat not far behind, shepherding the two cubs and preventing them from wandering too far from the adults. There was good reason for such protectiveness: there were caiman in these waters, and a giant otter cub that lost contact with its family group would prove easy pickings for these six-foot relics from the age of dinosaurs. Camoodies (anaconda), too, would find an otter cub extremely attractive fare. But within the confines of the group a cub seems relatively safe. One thing did puzzle me about this arrangement however: why was it, I wondered, that the giant otter is not attacked by piranha? Any human that spent as long in the waters of tropical South America as the giant otter does would certainly end up minus a few fingers, a hand, or worse. Yet the giant otter seems to glide through these dangerous waters without even a second thought for the fish which George Myers, a Stanford University ichthyologist, has described as being 'feared as no animal is feared throughout the whole length of South America'. But then, perhaps if the giant otter could talk we would hear a different story; no-one knows just how many giant otter *are* eaten by these demons of the river.

I felt very excited as I watched the otters nearing our hide-out. This was the first time they had bumped into us unknowingly, as opposed to our actively seeking them out. Now at last we could watch their behaviour with very little chance of their knowing we were there, for it was obvious from their actions that they had no inkling of our presence in the smaller creek. They were swimming steadily upriver, and I noticed that the adults swam with much more of their upper body showing above the water than did the cubs. The youngsters seemed to find swimming much harder work than their parents, and consequently floated lower in the water. Spotted Dick and Black Throat hummed or cooed occasionally to Ringo and Anonymous, as if to reassure them in this hostile world. The coo sound was especially interesting as it was new to us and could be added to the list of otter vocalisations which we were already familiar with. At that time we knew only three of the giant otter's calls, the commonest being a sharp 'hah!' sound, given when the otter was startled by some strange or potentially dangerous object. Snorts were given in this situation, too, especially if the animal felt itself particularly threatened. When it was really annoyed, the snorts graded into growl-barks, normally emitted when the animals were bluff-charging our canoe, as on the first meeting with the residents of Lama Creek.

To our intense pleasure, Spotted Dick suddenly stopped swimming, plunged down into the water next to the bankside and emerged with a fish some fifteen inches in length hanging between his dripping jaws. It looked like the long golden body of a hurri, a species which we knew spent most of its time lying in shallow water near the bankside, waiting for passing fish upon which to prey. Now it was prey itself, held just behind the head by Spotted Dick's strong jaws. The male otter was very excited by his good fortune and thrashed about in the shallows, shaking his tail wildly and making low, and to us very menacing, growls. None of the other otters approached him, seeming to be too intent on finding their own fish. As we watched, Spotted Dick seized the body of the fish between his two webbed forepaws and deftly re-positioned his mouth over the head. There was a sickening crunch as his jaws bore down on the hard scales of the hurri's head, crushing them to pulp. The fish's tail wriggled madly for a moment in response and then Spotted Dick settled himself down in the shallow water for a feed. Everything was consumed, the male otter taking the fish's head first and eating everything right down to the tail, guts and all, while throughout the feast giving voice to a low-intensity hum-growl as if warning the other animals to keep away.

This aspect of Spotted Dick's behaviour surprised both Liz and myself. We had supposed that so social an animal as the giant otter might indulge in food sharing, as do other social animals such as the cape hunting dog and the chimpanzee. But on the basis of this evidence, it seemed that in the case of the giant otter it was very much 'every otter for himself'.

The rest of the family were busily trying to emulate Spotted Dick's success and it was interesting to watch their fishing technique. Black Throat rushed about in the shallow water among the moca-moca, making a tremendous amount of noise and splashing water everywhere. The idea seemed to be to scare the fishes lurking in the shallow water into trying to make a run for it, then the otter would pounce and try to capture a fish as it fled. Since many of the fish which the otter eats live in holes in the bank and other hard-to-get-at places, scaring them in this way must be quite an effective ploy. The two young cubs were imitating mother, but watching their ungainly gambols through the shallows it was hard to believe that they understood the reason for this 'game'. I had the distinct impression that they were simply enjoying themselves. As they careered through the bankside vegetation, tripping over broken stems and barking excitedly, they looked the picture of happiness.

Their mood soon changed when mother finally managed to catch a fish. This time it was a therooma, a night-living catfish that lies up in the mud at the bottom of the creeks during the day. As soon as she caught her prize, the cubs surrounded her and with open mouths and heads held high proceeded to give vent to an awful, siren-like 'wah-wah-wah-wah'. On and on it went, seemingly without pause for breath, until, if I had had one, I would have been perfectly willing to part with my last fish if only the two little fiends would stop their hideous racket. But their mother was unimpressed and continued to consume her catfish at a leisurely pace without even a glance in the cubs' direction. She may have been threat-growling too, like her husband, but from where we sat the sound – any sound – would have been drowned out by that awful 'wah-wah-wah'. To our immense relief, the noise stopped as if by magic as soon as the last piece of tail was consigned to the female's stomach, and the two youngsters began their play-fishing again, though I fancied I could see a little more purpose in their activities this time.

One thing that many Guyanese had emphasised was the amazing clannishness of the giant otter. One chap, Mr Arnold

Correia, was in the habit of hunting in the East Demerara Water Conservancy which formed part of our study area. While pursuing tapir, he and his companions had disturbed a family group of six otters, two of the cubs being very young indeed and scarcely able to swim. By firing their guns in the air, the men succeeded in chasing off the four older animals and took the young cubs aboard their boat, intending to rear them as pets. The parents, and what were probably their two elder cubs, gave chase, but by making use of their outboard engine the men were easily able to outdistance them. They continued on their way for another seven or eight miles, finally making camp on a small island in the conservancy just before the sun went down. The two cubs were fed with condensed milk and a little raw fish. They were placed on some dry grass, and a wicker basket weighted with two heavy stones was put over them before the men retired to their hammocks for the night. About two o'clock in the morning they were awakened by an almighty rumpus, and the first man to discover his torch found himself looking down the beam at the rear ends of several otters who were at that moment making for the safety of the water. More torches flashed on and in their light Mr Correia saw six otter heads swimming rapidly into the darkness. The two cubs were gone. Later, the same giant otter group was seen in another part of the conservancy. Mr Correia did not try a second kidnapping; from that time neither he nor any of his friends has harmed a giant otter. 'If they think so much of their kids they'll travel eight miles to rescue them,' he told us, 'I think they deserve my respect.'

I had been snapping away quietly with the stills camera, wonderfully pleased with myself at getting such close shots of fishing otter, while behind I could hear the soft, incessant scratch of pencil on paper as Liz hurriedly scribbled down notes. Dimly, I was aware that the scratching had stopped, and a split second later Liz was screaming horribly. Snake! was my first reaction and dropping the camera heavily, I spun round in my seat, expecting to see a bushmaster or fer-de-lance impaling itself on my wife's arm or leg. There, on Liz's bicep, I saw a hideous, nightmare creature, at least six inches in diameter and completely covered with obscene black hair, from which peered six unblinking eyes. It needed only a second to take all this in, then Liz suddenly stopped screaming and came to life again, trying to brush the ghastly creature from her arm. Her first attempts failed, the spider merely pressing closer to her skin, but as I moved forward she finally managed to dislodge the brute. The spider hit the gunwale of our boat and seemed to rebound like a rubber ball straight between us, landing with practised ease among the river-side branches. In another instant it was gone.

Horror turned slowly to relief. There had been no real danger. Despite its off-putting appearance, Liz had merely been used as a tree-branch by a bird-eating spider, one of the largest of its tribe and a creature that makes its living by pouncing suddenly on whatever small creatures it happens upon in the forest. Lizards, birds and even small mammals like mice are all taken by this formidable arthropod, but humans are objects more to be avoided than actively hunted. Even so, I had to admire Liz for her courage and presence of mind; in common with many people, I have an unreasoning terror of spiders. There is something in the human psyche that detests these eight-legged horrors, and to find so large a spider on one's body, against one's bare flesh, would have been enough to send me mad. How Liz had managed only to scream I don't know. Had the positions been reversed, she would undoubtedly have found me in the bottom of the canoe, stone dead from shock.

The otters, too, had been shocked, and by the time we thought to look for them they had vanished. But which way, up or down river? Acting on a hunch, we moved away from the rest-house until we had paddled further down Lama Creek than we had ever explored previously.

As we neared the next bend we were attacked – there is no other word for it – by an overpowering stench, a mixture of otter, rotting fish and cow-dung. Breathing through our mouths to avoid the worst of the smell, we turned the corner and discovered the source of this unwholesome aroma. On the inside bank of the creek, about three feet above water level, was a huge, bare expanse of ground. Not a single tree, bush or blade of grass grew upon it, in marked contrast to the profusion of growth in the surrounding jungle. Immediately, the words of Sir Robert Schomburgk, who was writing 140 years earlier and was one of the first scientists to explore tropical South America, leapt into my mind:

> . . . their feeding places are so devoid of vegetation, if we except the larger bushes and trees, that they cannot be mistaken. . . . A complete path leads up to these places, which, in consequence of their ascending and descending in single file, is hollowed out . . . their haunts are easily known by a strong and disagreeable smell, in some instances so strong that we increased by all means in our power the speed of the canoes to get out of its precincts.

Schomburgk had been fortunate, for he had been able to escape the offending odour as quickly as possible. Our job was to make straight for its source, to land and to investigate this otter 'camp site'. The pong was so all-pervading by the time we beached the canoe that I felt almost physically sick. Liz began the preliminary data-collecting by measuring the site: length, width, number of entrances, distance from the river. As Schomburgk had described, the camp site was perfectly bare, almost as if some proud hausfrau had just swept the hard-packed earth clean with a vacuum cleaner – except for one spot. About three feet from the farthest end of the site was a huge pile of scales and bones from various fish species. We examined this area and found it to measure three and a half feet in diameter (compared with forty-three feet and eleven feet for the length and breadth of the site respectively). I pushed my knife into the scale mound and discovered that it was at least an inch deep in several places. Someone had been having quite a feast here over a long period of time!

Liz took a sample of scales from the thickest part of the mound, using a six-inch-square grid that we had brought with us. She did this at all the marking sites we discovered during the course of the study. The freshness of the scales would give us some indication of how often or how long the site had been used. We also planned to create our own otter site. Then, every month, we could return to our counterfeit site and see what plants had re-colonised the bare area, keeping a record of growth until the site had been completely grown over. This would allow us to estimate the length of time elapsed since a site had been abandoned. Using this small battery of tests, we could discover a lot about our subjects without even seeing them. We could tell whether giant otter were in the area, how long or how regularly they had been using their camp sites, or, if they had left this stretch of river, how long it had been since they disappeared.

The scale sample Liz took had another use. We knew that each of the scores of fish in the Guyanese rivers had a distinctively shaped scale pattern. But the scales in our sample comprised only six or seven different patterns. If we could identify the owners of these scales we would have a very good idea of the favoured prey species of the giant otter in this area. Liz spent hours washing, drying and sorting through her scale samples. She used a low-power microscope to identify the smaller fragments, some as tiny as pinheads. But even with the help of these optics she seldom found many of the small translucent scales that would signal the presence of a fish called the lukanani or peacock bass. This beautiful, black-and-gold barred monster (it can grow to a length of three feet or more) is highly prized as a fast, hard fighter among Guyanese sportsmen. Many of these men had told us that they would have no compunction in shooting giant otter whenever they came across them because they were certain that they ate huge quantities of lukanani and so spoiled their sport. We believed that this story was nothing more than a fable and that the situation was similar to that in Europe, where an almost identical belief existed. There it was almost an article of faith among anglers and gamekeepers that the Eurasian otter, *Lutra lutra*, killed and ate salmon. As a result it was greatly persecuted. Then someone actually looked at otter behaviour and showed that most of the time otters hunted slow-moving coarse fish such as chub or dace. By showing that the lukanani was only rarely taken by the giant otter (and both the scale samples and our observations pointed in this direction) we hoped to save the river wolf from much unnecessary persecution.

Liz was glad that she had demolished this misconception, but convincing the old school of rod-and-tacklers proved something of a battle. Their expressions said everything: here was a mere slip of a girl who knew nothing whatever about fishing telling them, veterans of scale and fin, what ate or did not eat their prize fish! It took many friendly exchanges, countless rums and a few peers through the microscope to convince them that they had, perhaps, been a little excessive in their accusations. 'The water dogs probably scare the living daylights out of the fish,' Liz conceded diplomatically, 'but it's only temporary. If you let things settle back down after they've gone, I bet you'll be back in business in no time.' After that, we were pleased to notice that one particular sportsman stopped threatening to turn 'those damned water dogs' into bedroom slippers. The information had been absorbed but grudgingly: it meant he would have to find another scapegoat for his less than perfect fishing technique.

The otters themselves certainly had no worries over technique. Theirs was an inborn ability honed to perfection by years of practice. And they had no-one to blame if they missed their fish. Over the weeks Liz gradually worked out their main hunting methods: mid-river hunting and shallows hunting. Both approaches were used in the creeks but in Russel Lake, where virtually all the banks were steep-sided, there was very little shallows fishing. Mid-river hunting required deep vertical dives that usually lasted about forty-five seconds though sometimes they could go on for as long as two minutes. The otter would spear its head forcefully into the water, bring its back, rump and tail into a sharp arc and kick out with its back legs. Thus fuelled,

45

the body jetted downwards into a steep U-dive. This sort of diving was quite different from the smooth sequence of shallow dives used during travelling. For this, the otter pushed its head into the water at a shallow angle, gracing us with a glimpse of silvered shoulder. Its back and tail followed through underwater, quite unlike deep diving where these bits of anatomy could be seen clearly from above. Travelling dives lasted about ten to twenty seconds and alternated with five-second breaks at the surface.

Mid-river hunting was fairly well synchronised among the family members. Sometimes Black Throat and the two sub-adults dived together, followed a second or two later by Spotted Dick; sometimes it was the other way round. Occasionally, the dives were separate but in rapid sequence. Liz was keen to know whether this harmony signified a form of co-operative hunting. Several fishermen had told us that large groups of giant otter fan out into a skirmishing line across the width of the creek and move the fish slowly before them, rather like beaters on a pheasant shoot. Whenever a fish, alarmed by the number of otters and the commotion in the water, breaks cover and tries to escape, there is always an otter nearby to pounce on it. This sounded a rather plausible story. Herskovitch, a naturalist of the last generation, also claimed that giant otter families hunt together as a co-operative group, driving fish into a shallow backwater where they are easy prey. But plausible or not, there was certainly no indication of such behaviour at Lama. We never saw family groups herding fish into the bank and nothing the animals did suggested such an intimately integrated strategy. The harmony of movement in mid-river hunting seldom involved the formation of a skirmishing line and, as we had seen with Spotted Dick, all fish were eaten on the spot by the individual that caught them. Not so much as a fin was shared with the other family members.

It was only after several months that we realised that the giant otters' failure to co-operate with each other could be fairly easily explained. Well-co-ordinated pack hunting is usually seen only where the predator is smaller than the prey it hunts, as when cape hunting dogs pursue a wildebeeste. In the otter's case the fish were always much smaller – never more than a third of a single animal's body length.

What, then, was the point of fishing together as a family? An encounter Liz had with a single otter in Russel Lake suggested an answer. It was hunting close to the sedge, and Liz timed each dive the otter made and noted how many fish it managed to catch. Not only were the dives longer than those of individuals in a family group – about one and a quarter minutes per dive – they were also less successful. Only about one in twenty dives produced a meal compared with about one in ten to fifteen for a group member. It seems that if you are a part of a family group you stand a much better chance of a full belly than if you have to fish alone. Perhaps, Liz reasoned, group hunting in deep water made individual catches quicker and easier because the greater number of animals caused a general panic among the fish population.

Shallows fishing was even more efficient than group deep-water fishing. It was also a lot noisier. There was a strong element of concentration and tense excitement that was totally absent from the silent competence of deep-water hunting. Strangely, group movements were always far less cohesive during a shallows hunt. Family members became separated from each other by as much as forty feet and Spotted Dick especially tended to lag behind the other three. Too often, the thick palisade of moca-moca frustrated our observations and it was a relief when the stalks thinned and we could get a clear view of what was going on.

We had seen that the cubs were not very successful when they first tried shallows fishing for themselves. As four-month-olds they produced more splash and squeal than fish, but their technique did slowly improve over the months. After a year of ham-pawed trial and error they were as skilful as their parents in the rush-and-pounce routine. Roughly three fish were caught every hour by each otter, compared with only one or two for mid-river hunting. Why then did the giant otter remain in such large family groups? It did not seem likely that deep-water fishing was the only advantage that accrued from such a social structure. Could the answer lie in the camp sites the otters constructed?

Although they had provided us with much information, the scale and bone mounds of the camp sites were still something of a mystery. Some workers maintained that these monumental remains were latrines where the giant otter family 'sprainted' (defecated) communally as a form of territorial behaviour. Others claimed that the mounds were dining areas, cafeterias where the giant otters came to eat after catching their prey. This too seemed a valid explanation. So what were the scale mounds, the remains of an otter feast, or of otter faeces?

We inclined to the latrine theory, for, as we knew to our cost, the sites' aroma was certainly not of roses. One fact did puzzle us, however. On examining the scale piles, we discovered numerous intact catfish spines. These spines, found on the pectoral and dorsal fins, are held erect by the fish whenever danger threatens. Not only are they pointed at their tip and remarkably strong, but running down each side of the spine is a saw-tooth of short, razor-sharp needles. The spines are certainly very effective against man – on several occasions I had hooked catfish while fishing and the spines, one on each 'arm' and a third just behind the head, had made it extremely difficult to handle my catch with safety. The Amerindians claim that the spines are poisonous in some species, which makes the fish even more troublesome. It is amazing that the giant otter can totally consume these catfish and have the spines travel the whole length of the gut without suffering any trouble from punctured stomach or intestine; it needs only a slight puncture of the gut-wall for infection to set in and kill the animal. The answer probably lies in the large amounts of mucus that cover the otter's alimentary canal and so help to protect it against abrasion from this fishy equivalent to roughage in its diet.

In plumping for the latrine theory we were going against Guyanese folk-tales on the subject. One old fisherman told us that giant otters not only fish communally, they eat together too. This was a very suspect suggestion, though the old chap got quite indignant when we questioned its veracity. Still, it did help to explain those enigmatic mounds of fish-scales. According to the fisherman, a 'big wata dog', after catching a fish on one of its communal fish-beats, will carry the fish to the bankside and leave it on a pile caught by other otters in the fishing group. One of the otters sits by this fish-cache as a sentry, to protect the spoils from tegu lizards, black vultures and caiman, any one of which would be only too happy to avail itself of such an effort-free meal. When the hunt was over, all the otters were said to return to the food-pile and to share the catch equally. If this were true it would have a profound effect upon our view of otter behaviour – altruism on such a scale was extremely rare in the animal kingdom. And what a sequence to photograph! But even better was to follow: with his hand on his heart, our informant claimed that on one occasion, while the rest of the otter group were away fishing, he had seen the 'sentry' sneak two fish from the pile and eat them. On their return, the rest of the group jostled around the catch, seemed to count the fish and, realising their loss, set upon the sentry. 'Oh Lord!' The old man struck his thighs and laughed at the memory. 'Man, dey push he in de water and beat he bad!'

Such a sequence would be a delight to capture on film, a real 'nugget' in cameraman's parlance. We would often fantasise on such a triumph, but the reality of our situation was unfortunately a little more prosaic. The giant otter was proving as elusive a subject as the most camera-shy VIP.

5 Otter Detectives

(Liz)

Although the study side of the project was coming along nicely, the same could not be said for the filming. Apart from the difficulty of finding our giant otter quarry, to photograph them successfully they had to 'perform' in bright sunlight. This they seemed unwilling to do and I would often see Keith's face set in disgust as Spotted Dick and his family sported and squealed among the dim gloom of moca-moca fringing the river bank.

When they did venture out into the full sunshine there were other problems to contend with. Our canoe, despite its flat bottom, was by no means the stablest of camera supports and the slightest movement produced an unacceptable amount of camera-shake. The narrow boat also meant that Keith was confined to a single line of fire – the canoe had to be pointing directly at the animals before he could see them in the viewfinder. If the otters moved, which they always did, I had to keep turning the boat towards the target in order for Keith to follow the action. As soon as he tried to swing the lens independently of the canoe, he 'lost the horizon' and the film would show the world slipping either to the right or to the left. We quite often ended up quarrelling because it took a great deal of concentration and physical effort on my part to keep steering like a guided missile, and I wanted to observe and take notes at the same time. To make matters worse, we sometimes misinterpreted each other's hand signals in the heat of the moment. Our otter friends seemed to take a delight in zig-zagging across the river, diving in unison just as Keith had focussed on them but before he could press the button. We had to find an answer. Something had to be done to improve our chances of getting good footage and at the same time allow me to observe the animal I had come to study.

The solution came by trial and error. At first we debated whether we should hire a boatman to steer us but even if we had had sufficient funds, it still would have done nothing to reduce camera-shake. We tried to stabilise the canoe by tying an empty oil-drum to either side but I found it impossible to steer, so if anything the turning problem was worse than before. For a time, we actually abandoned filming on water and took to lying in wait on the more solid river bank for our otter family. This strategy was doomed to failure. It works beautifully for mammals with small home ranges who do the rounds of their property frequently, but with a creature as wide-ranging and unpredictable as the giant otter it would have meant weeks in one spot twiddling our thumbs. No doubt the sit-and-wait tactic would have paid dividends eventually, but we did not have ten years at our disposal! Following our quarry by boat was the only way of maximising filming and study opportunities.

Three weeks were to pass before Keith hit upon the answer. It minimised camera-shake and gave me time to tape-record my observations. He made a monopod from one of our small folding tripods. By carefully adjusting the tension in the ball and socket head and by holding the whole apparatus firmly between his legs, he could compensate for any boat-shake and keep the horizon reasonably straight no matter which way the boat turned. The husband–wife rapport improved considerably after that. I no longer felt like throwing my paddle at Keith's head and I suspect he felt much less inclined to chuck me overboard. A mutual tolerance settled upon our mission.

We searched far afield for signs of other giant otter families, investigating every likely marking site and pricking up our ears at the sound of every snort or sneeze-grunt. Most mornings would see us pushing off from the boat-house, our canoe laden with camera equipment, tape-recording gear, notebooks, food and machetes. Hats were imperative, for without them we would have been baked to a frazzle under the 120°C sun. Even so, I often felt on the verge of heat-stroke by mid-afternoon, having motored some fifteen to twenty miles and investigated as many promising inlets. The days slipped by in a series of long, hot flashes but I was gradually beginning to find out more about giant otters.

For one thing, they were extravagant architects. Over the weeks we found dozens of their monumental marking sites, some fresh and smelly like the one we had first discovered, others old and derelict but still quite bare around the latrines. Most of the eighty-odd sites were located in Russel Lake on the small islands and containment dykes. Only thirteen of them had been made on the forested banks of the adjoining creeks but these sites satisfied the same basic building requirements as those on the lake – plenty of shady tree cover and a nice flat, unfloodable surface. It did not matter what type of tree provided the shade so long as the site was kept cool. Jamoon, bamboo, Ite palm, mock paw-paw and fig all did the same job equally well, none of them being preferred over the rest.

I plotted all the sites on a map of the area and found to my surprise that they were distributed in distinct clumps. The clumps covered an area of about eighty square miles and no doubt included the home ranges of several family groups. It seemed a rather odd thing to do, to mark one specific area of your territory intensely and leave large gaps unmanned: not the sort of defence strategy a self-respecting otter should employ. Miles of shady, dry land lay unexploited between the clumps with only the odd site dotted here and there to show its ownership. Perhaps the site clusters represented vulnerable spots that were particularly susceptible to would-be invaders, though this did not appear to be so. I concluded for the time being that they probably reflected some aspect or other of giant otter sociality. Marking sites were after all more than military outposts. As bearers of urine, spraint and scent, they were to giant otters what lamp posts are to dogs: olfactory answer-phones. They communicated long-life inform-ation on the identity, sex, age, mating receptiveness and health of the depositor as well as the time elapsed since marking.

Other issues on territory began to emerge amid the cross-fire of query and counter-query. It was early days yet, but it seemed to me that Spotted Dick and his family were not all that bothered about defending their home range. Carnivores are well known for their flexibility in the property-holding business. Some en-vironments encourage fierce territorialists, while in others the order of the day is tolerance, and this wide spectrum of behaviour can be found in one and the same species. Giant otters, then, could just as easily range freely, and prevent conflict by mutual avoidance, as defend a specific area. I was aware that this trend of thought was quite contrary to what Nicole Duplaix had found in Surinam. Her five family groups were hard-line materialists, defending well-defined territories along a nine-mile stretch of Kaboeri Creek. Trespassing was uncommon and when it did occur, the guilty animal looked decidedly nervous and hadn't the courage to venture too far into the alien territory. For my part, I had not observed any such cut-and-dried allocation of real estate in my Demerara family group. If anything, they appeared nomadic.

I checked most of the marking sites every week or two to see if they had been visited. The time came to examine the large group of sites on the nine-mile-long east dyke of the conservancy. Afterwards, we planned on cutting across the lake and inspect-ing a spit of land called Anadale Gutter which the men reckoned

51

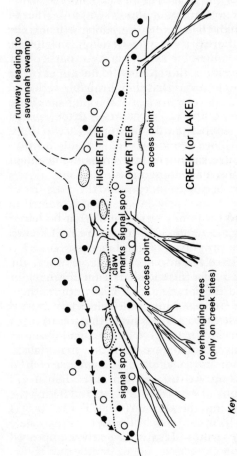

Key

○ trees 3–12″ in diameter at base and 20′–60′ tall

● saplings (creek sites) or grasses, heliconias etc (dyke sites)

⬭ family latrines or spraint heaps (up to 5 per site but only one used at a time); remains more dispersed on dyke sites which are more open to rain and wind

→ site extension, which occurs on some sites

········ 1–2 foot 'step' between lower and higher tiers (only on some creek sites)

A typical giant otter marking site

had 'plenty, plenty' otters. The day promised to be hot and humid, which was surprising as it was mid-September and well into the dry season. I wiped the sweat off my face and tried to push the muggy discomfort from my mind. Our canoe chugged past mats of water lilies and floating weeds, causing loud panic among a female jacana and her three chicks. A fish suddenly flew out of the depths beside us, ski-ing along the surface for a few yards before flopping back in. The vast brown velvet of sedge extended in all directions as far as the eye could see, fusing hazily with the horizon. Through the smudge I could just make out distant groves of Ite palm, sticking up like needles in a pin-cushion.

By one o'clock we were about to make for the shade of an overhanging jamoon tree when we suddenly caught a whiff of otter. I immediately cut the engine and we paddled quietly towards the source of stench. On the bank ahead of us an otter was 'dancing' across the newly trampled earth, swinging its tail and nether regions from side to side and scraping the clay with its forepaws. Trembling with excitement, we stopped the canoe behind the leafy branches of a young jamoon tree and started filming. My diary notes describe what we saw:

> It's a female, fully grown with a large, unblemished bib of cream. She continues her animated floor show for a full two minutes, pulling down grass stalks and low bushes and rubbing them into the mud with rapid, circular movements. She stops and waddles over to a small blob of spraint in the middle of the site and sniffs it repeatedly. With her back to it, she suddenly raises her tail, hips arched, legs stiff and webs slightly out-turned. Her back legs lift alternately like a duck marching time and a spray of urine falls just behind them followed by a long cord of dark spraint. She waves her tail up and down four or five times and matches the movements with a few head bobs. Without actually trampling the spraint, she pats the mud around it and a minute later resumes her plant-rubbing operation.

How long she would have remained on the site we were never to know because at that moment one of the camera lenses rolled along the wooden seat and startled our performer. She gave a loud snort of alarm and leapt into the water, disappearing in a comet's tail of bubbles.

This was a unique sighting. Nicole Duplaix had never once observed an otter marking by itself, it had always been a male—

female pair. We searched for a possible partner in the surrounding water but found nothing, and when our dancer did surface a couple of hundred yards away, there was no other head alongside hers. Knowing how inseparable giant otter mates are, it was a fair bet that this female was a spinster, albeit a very young one. She may have been setting up a territory and at the same time leaving a coy invitation for passing males: I am sexually mature and ready to pair. On the other hand, she may actually have been responding to a deposit from a male who had visited the same spot beforehand. Whatever the reason for 'Dancer's' behaviour, it was something I had not expected to see, a bonus out of the blue.

Before pushing off for lunch, I climbed onto the site for a quick look. There were no other sites nearby; this was one of the few totally isolated ones. Most of the dyke's twelve-foot width had been bared of vegetation and long strands of abrasive cutting grass, moca-moca and flowering heliconias lay buried in the soft clay. Claw-marks scored the entire piece of otter sculpture and a teaspoonful of green ooze indicated where the animal had ejected its scent in fearful response to the noise of our camera lens. Otter scent normally coats the spraint in small quantities, but at times of intense excitement or fear it is squirted out independently. More than just an erotic perfume, it is the magic substance that gives spraint its powers of communication. The two spraints we had seen earlier were still dark and wet and covered in mucus. I weighed and measured them, putting half of each in my sample bags for later analysis. The fact that they had not been trampled was rather strange but I was to find many more like them later on in the study.

By three in the afternoon we had reached Anadale Gutter, the unsurveyed canal about seven miles into the swamp. Here I discovered my longest otter site. It was not a recent one – obviously it had not been visited for several months – but I found four latrines spread out along its eighty-foot length. As on most other sites, several mature trees provided shade for the greater part of the day and I counted three signal spots, each a small scoop in the ground which the otters had dug with their forepaws. We often found these on communal sites as well as next to single spraint. They acted as a visual supplement to the otter's olfactory message of spraint, scent and urine. According to the Swedish zoologist, Sam Erlinge, the European otter, *Lutra lutra*, also sculpts signal spots but it leaves sign heaps as well. These usually take the form of tufts of grass that are twisted into a 'tower' by the otter and on which it deposits a single spraint or blob of scent. We never found our giant otters to make anything like them. They

54

are probably characteristic of *Lutra* otter species if not of *Lutra lutra* alone.

By dusk, we had checked a total of forty-two otter sites and put another seven new ones on our map. Add to that our privileged peek of an otter actually making a site and our satisfaction at the day's work was fully justified. Night fell quickly as we slowly motored back to camp but there was a moon so we weren't particularly worried about hitting floating islands or getting lost. Besides, the large white lilies that lined the open waterway stood out like cats' eyes against the dark water and guided us away from the reeds and hidden logs. We relaxed and took in the stars: the Seven Sisters, the Plough, Orion and the white mist of the Milky Way. There were no city lights or buildings to blemish the experience and looking upwards, it was as if the two of us were with the stars, suspended in space. There was no breeze to disturb the stillness and even the mosquitoes left us alone. I was so immersed in my role as a heavenly body that I thought for a moment I had seen a spaceship. A dark shape flashed over us and disappeared 'offscreen'. It returned and this time I followed it back down to earth or, rather, water. 'Fish-eating bat,' said Keith. 'Did you see the size of the fish it got?' It came in again, skimming the surface of the water. There was a small splash as its long thumbnail – a feature unique to *Noctilio leporinas* – speared the prey, and pulled it into the air. Not for the first time I envied all winged creatures their mastery of the air.

The ability to fly would have helped me enormously with the study. After five months of fieldwork, Spotted Dick and Co. were still not as amenable to direct observation as I would have liked. I actually contemplated hiring a balloon and spying on them from above but balloons are the servants of thermal currents and capricious breezes and I doubt if I would have been able to keep course. I finally decided to try a more mundane tack. I would estimate when Spotted Dick and his family were due to visit Dark Lane (I was beginning to find that they visited the area fairly regularly) and paddle down Lama for four or five days at a stretch. If we glimpsed them in the distance coming upriver, we would then heave our canoe into the bushes and wait for them.

I set off to test the technique myself one morning at dawn. The hoatzins were still asleep as I slid the canoe into the open creek but a group of black-crown night herons took off with a rush when they saw me. The sky wore a bizarre face that morning. Rain clouds darkened the eastern sector but to the west the sky was a clear slate-blue. It was as if the two poles had been reversed and the sun had decided to rise in the west. As I watched, the black ceiling ahead slid slowly westwards but long, wispy tendrils clawed at the orange dawn as if loath to leave it. For several minutes it smouldered and scratched at the central glow but the pull of the main body of cloud was too powerful. It finally relinquished its hold, bringing the rain towards me. I got soaked, but it proved a good omen.

Far downriver where the torrent had passed and the sun was now warming the water, I could just make out a silvered ripple. My binoculars picked out two giant otters. Another pair of heads surfaced beside them and the foursome then swam together for a short distance in my direction before diving in unison, backs humping with the effort. They had not seen me and I steered the canoe into a shallow bay in the river bank where the drooping branches of a half-drowned fig tree provided good camouflage. I had estimated from previous observations that, when fishing, giant otters travel slowly, about 1 mph, quite in contrast to the 4 mph average of surface swimming. They can move much faster when totally submerged and I had recorded underwater speeds of up to 10 mph for short distances. In fact, this rapid movement was their favourite method of escape from prying zoologists; a fast swim underwater could take them several hundred yards from our boat in a few seconds without once coming up for air.

On this occasion, Spotted Dick and his family were quite oblivious to my presence and continued to fish leisurely in midriver. I took the opportunity to time their dives and to note the success rate of fish capture. I saw only two successful catches – one each by Spotted Dick and Black Throat – in the three minutes it took for them to reach my hiding-spot.

Silently they worked their way towards me and were now close enough for me to hear the soft explosions of air as they surfaced from each dive. The paddle-and-hide ploy had worked! I had managed to collect some good fishing data but there was something else I noticed as well: on the basis of head size, I could no longer tell Spotted Dick and Black Throat apart from their two cubs. They had obviously done a lot of growing since our first encounter seven months ago. Then, their heads had been about two-thirds the size of their parents and I had judged them to be about four months old. But the little water dogs had now reached adult size, though this did not necessarily mean they had also become sexually mature. I noticed, too, that the family no longer formed the old two-plus-two formation of parents in front and cubs behind. All that had disappeared. They still kept fairly close together but I could no longer detect any set arrangement. Perhaps now was the time to find out the chest pattern of the second cub. We had Ringo's cream collar but a question mark for her brother, Anonymous. Before, any attempt to get him to periscope out of the water had failed but there was a good chance that the confidence of adulthood would stir a response. I pinched my throat between thumb and forefinger as an old Amerindian fisherman had taught us and 'caterwauled' long and hard. It was an imitation of the giant otter's frustration call and I found it as therapeutic as they probably did.

Up bobbed a chest: it was Black Throat. Spotted Dick rose up beside her, or so it seemed at first glance, but it wasn't Spotted Dick. It was Anonymous, proudly exposing a mottled chest pattern similar to his father's. I re-christened him Spotted Dick Jr. Behind them, Ringo made a half-hearted bid to periscope while Spotted Dick himself charged in, snorting, from the side. They had all turned towards me – just what I wanted – and peered earnestly through short-sighted eyes. Spotted Dick Jr continued to bare his chest and kept attacking and retreating. He had certainly undergone a big character change since his juvenile days, revealing a bold streak I had never thought existed. If I was not mistaken, he had actually eclipsed Spotted Dick's role as group defender. Either that, or he had let his adolescent fervour get the better of him.

The family was still in the area the next morning. They had spent the night in Dark Lane, as was their custom, and were now doing a spot of early morning fishing in the headwaters. But on my third day out they were nowhere to be seen and I reckoned they had travelled on to another fishing haunt in their home range. I paddled the entire five-mile length of Lama, to the point where it meets the Mahaica River. Exhausted, I clamped the engine to the hull and motored back upriver, thankful for the rest. Groups of gossiping anis took off, screaming, as the boat rounded each bend, the sun catching their feathers and turning them into brilliant flashes of blue-green iridescence. I was within ten yards of Dark Lane when I saw the otter. I cut the engine immediately and pulled the boat into the left bank. It had seen and heard me but it did not periscope in curiosity as family groups did. Instead, it continued to swim quickly past in a downriver direction, alternating surface swimming with long stretches of underwater travel, clearly wanting to get somewhere in a hurry. It did not allow me to get a glimpse of its chest pattern but it was certainly an 'incomer', a trespasser in Spotted Dick's territory. I wondered what would happen if they met. More than likely, both sides would try to avoid a meeting like the plague, avoidance being a lot less wasteful than conflict. In the European otter, resident adults tolerate loners like this one and actually allow them to live temporarily in their territories for part of the year. The aliens are then called temporary residents but are forced to become transients or passers-through for the rest of the year when competition for food and for mates increases. Whether a similar relationship existed between giant otter family groups and single otters was quite unkown but I felt that the loner I had just seen was a transient. It certainly looked nervous enough and moved swiftly, not dallying to fish on the way. The female we had seen 'dancing' on the east dyke a few weeks ago was rather more of a problem. She had advertised her presence in no uncertain terms which indicated some measure of status. Perhaps she had been on safe ground not owned by anyone in particular.

I checked the marking sites in Dark Lane to see which one Spotted Dick and Co. had used during the night. Judging by the immense growth of moca-moca at the access points, it was a site they had not used for close on two years. But it was not this that excited me. Far from the main latrine was a freshly deposited single spraint, something I had never found before on a communal site. There was another one on the group's favourite site on the opposite bank. At the time, I did not see the connection between the single spraints and the transient. A lot more data had to be collected before the link became apparent, but when it did, the relationship between transient and family group finally became clearer.

We were shortly to be told of solitary otters of another kind. As we motored back to the rest-house one evening, we were ahoyed by someone chopping wood on Russel Lake's east dyke. It was Coxy, the watchman in charge of the sluice gates at Flagstaff. He had left his son to man the post, but as it was getting late he asked if we would give him a lift. 'Come and spend de night wi' me,' he grinned amiably, gold teeth flashing. 'Ram caught a big labba last night and we going to cook up a leg tonight.'

We had never tasted labba before but had been recommended to do so by many Guyanese friends. In fact, there was a saying in Guyana that if you ate this rabbit-like animal and drank creek water, you would always return. The brown water we drank pint by muddy pint certainly did nothing to substantiate the claim, and it was home to so many gut-loving bugs that we would be lucky if we managed to leave Guyana in one piece, far less return! Still, ever the optimists, we were ready and willing to test the meat part of the maxim.

The sun had slipped below the horizon by the time we reached Coxy's place, a small hut made of greenheart wood. A seine net was draped over an open shutter and a row of salted fish were curing on the palm-thatched roof. Apart from two slender paw-paw trees, the area around the hut was completely bare. To the west, the northern dyke stretched into a wedge-shaped perspective for more than fifteen miles behind the hut and to the east rolled mile after mile of savannah swamp.

Four dogs and three puppies rushed out to greet us, barking madly and causing Coxy's pet parrot to squawk convulsively. Ram Raj, the hunter, and two of his helpers were there to share the feast with us. A huge pot of black peas and rice had already been cooked on the mud stove and the labba was just about done. It turned out delicious, somewhere between chicken and pork. Fingers and chins glistened with grease and the shadowy light of the hurricane lamp made us look like primeval cavemen feasting on the day's catch. The statutory bottle of white rum was passed around. It anaesthetised my tongue but loosened Ram Raj's. He talked non-stop about all the animals he used to catch. 'Big, big cuffom fish, so long.' He extended his arms in typical fisherman style. 'Dey taste sweet, sweet but all gone now, no more see them. Nor de big water cow. Still get plenty of bush cow and labba, though. And plenty of parrots which mek good pets.'

'De twa-twa also good fo' pet,' butted in Coxy, pointing to a small russet bird in a cage by the window, 'but dey hard to catch.' Twa-twas, Coxy informed us, were a kind of passerine bird used in singing competitions. The sport was very popular and some men would pay half a month's wage for a good singer. The birds were judged on how long they could sing without stopping and because this ability depended on the size of their territories, the owner could increase his chances of winning by regularly taking his bird on long walks. This explained what Keith and I had seen some weeks ago in Georgetown: a man walking round the perimeter of the Botanical Gardens with a bird's head poking out of his shirt pocket!

Ram Raj was still talking as if he had not been interrupted. I caught the word 'otter' and pricked up my ears.

'. . . but I don't catch de small water dog as dey not good fo' eat.'

It took me by surprise. 'You mean a young giant otter?'

'Na, na, mistress. Dese be small even when grown up and dey not got white patch on chest like de big ones. De face be smaller, too, and de tail round, not flat like de big water dog.'

It sounded very much like *Lutra enudris*, the Guyana otter. A species about the same size as the European and North American otters, it occupies a small part of the giant otter's range, encompassing Venezuela and the Guyanas. Having been decimated by the fur-traders in the 'fifties and 'sixties, it was added, along with the giant otter, to the IUCN's list of endangered species. As part of my study I was keen to know how the two species managed to live together, but after nearly a year of studying giant otters at Lama, I had not once seen any evidence of *Lutra enudris* in the vicinity. The little and large duo did not live side by side in those parts of the creeks and swamp I visited regularly. There, *Ptero-nura* reigned supreme, holding exclusive rights over fish and holt. But it turned out that I had not looked far enough, though who would have thought of checking the rice paddies and plantation canals far to the north?

I turned to Ram Raj. 'In the rice fields? Are you sure, Ram?'

'Yes, mistress! True, true. De ranger of de Rice Board tell me he does see dem during de daytime. He often comes across two hunting down by Shank's Canal near de big sluice gate and also round Shank's Island. Dey much smaller dan de big ones,' he repeated, 'and lighter all over. You does see dem round here, too.'

An otter living in such an artificial habitat where sugar and rice crops were reaped three times a year! It was a most unusual find. I was raring to investigate the area for spraint and other signs and to see these little otters for myself. But unfortunately, Keith and I had to go into Georgetown the next day to organise our trip to the interior. We wanted to survey the giant otter and to compare its density there with the Lama population. Who could tell – our study site might not be the best place for studying or filming giant otters. The search for *Lutra enudris* would have to wait until we got back from the interior. With several hundred hours of field-work behind us we knew how to track and observe giant otters as well as any hunter. We had also learnt something of their ecology and marking behaviour and were now ready to put our knowledge to use.

6 Widening the Search

(Keith)

'Just watch this baby while I clean out her room. She's very sensitive, so don't make any sudden movements. Talk gentle, and remember the Golden Distance.' With these brief instructions, Martin van Hoof, the Snake Man, walked off to the far corner of his workroom and left me holding the 'baby' – a five-foot bushmaster. We were back in Georgetown en route for the interior, and in return for all the help he had given us, I had offered to photograph some of Martin's little pets for a book that he was writing on South American snakes. Already I was beginning to regret the offer.

The bushmaster, a favourite of Martin's which he insisted on calling Delilah, was lying curled malevolently on the top of a high bench. For protection I had a thin, three-foot-long aluminium snake-stick, the last four inches curved at right angles to the main body of the shaft. To my fear-filled brain it was totally inadequate, especially as Delilah was not satisfied with simply lying on the bench giving me snake-eyes. She was quite intent on making her way from bench to floor and heartily resented any attempt to hold her in one place. In reply to my half-hearted, ineffectual prod with the snake-stick, she raised her head malevolently, eyeing me coldly with an unblinking stare and silently shaking her tail. In her normal habitat among the leaf-litter of the forest floor this action would have produced a warning rustle every bit as effective as the rattlesnake's rattle, giving the bushmaster its alternative name, *la cascobel muda* – the silent rattle. Leaf-litter or not, Delilah was obviously not in the best of moods. It was a frightening situation and I silently cursed Martin – working unconcernedly in the far corner of the room – for putting me in this position.

The bushmaster (*Lachens muta*) is a pit viper and the largest venomous snake in South America. Lengths of more than twelve feet have been recorded, though ten feet is probably about par. At five feet long Delilah was only half-grown, but she was no less dangerous for that – the bushmaster's fangs and venom glands function soon after it hatches and even a youngster can inflict a mortal bite. This species has the longest fangs of any venomous reptile, almost one and a half inches in length and far longer than any New World snake (rattlesnakes, for example, never exceed seven-eighths of an inch). Like all vipers, the bushmaster has hinged fangs, which swing into their biting position only when the animal's mouth is fully open – in fact, the bushmaster's fangs are so long that it could never close its mouth with the fangs in their attack position. Backing up this formidable armament are huge venom glands, producing the poison which the bushmaster injects into its victim down the central canals of its twin hypodermics.

Delilah was on the move again. I made a quick jab to keep her busy and risked a hurried glance over my shoulder. Martin was not even watching but seemed instead to be completely absorbed in his task of cleaning Delilah's home, a six-foot aquarium. Though he gave no sign, I was sure he was enjoying every moment of my discomfiture. Then my eyes were back on Delilah. She was half off the table! I went into the attack again, slipping the curved head of the snake-stick under the bushmaster's body and flicking it as gently as I could back onto the centre of the bench. Throughout the action, I tried to keep at least the 'golden distance', as Martin called it, from Delilah.

According to the Snake Man, the golden distance is the closest a person can safely stand to a poisonous non-spitting snake with absolutely no risk of being bitten. It is simply calculated: the golden distance is exactly half the total length of the snake. It seems that snakes do not have the muscular wherewithal to cast all their bodies at an opponent or prey; they must leave some of their length on the ground to give purchase for the strike. Half their own length is apparently the minimum length needed to give this support. With a five-foot bushmaster (and a golden distance of two and a half feet), my three-foot snake-stick should have made me feel invulnerable. It might, had Martin not told me of an exception to this golden rule. The tommygoff (*Bothrops nummifer*) is a short, stocky cousin of the fer-de-lance, with an extremely short tail. Its strike is so violent that it is able to throw itself bodily at its victim, tail and all. That piece of information left me wondering if other snakes might also prove to be exceptions, Delilah included.

As she flopped back onto the centre of the table, Delilah was really angry, and she made a threat-strike in my direction before moving off again for the table-edge. That was enough for me. Let Martin catch his own damned snakes, I thought; all I'd come here for was photography. As I backed off, there was a blur of motion on my right and there was Martin, a second snake-stick in his hand and the bushmaster pinioned to the table, the stick pressing gently but firmly on the snake's spine just behind the head. Martin laughed with pleasure. 'Got to be fast with this baby,' he said, expertly bringing his free hand up to the back of the snake's head and gripping at the base of the cranium. Holding the head firmly between thumb and forefinger, he kissed the snake between the eyes. 'There, there Delilah,' he cooed to the scale-covered face, 'I gonna take you home.'

With Delilah safely ensconced in her newly cleaned room, Martin led the way up the bamboo ladder that formed the only connection between his snake rooms and the living quarters above. Earlier that day I had set up a small portable studio on the top of a large, low table in Martin's sitting room. It remained only for me to add two large stones to the set, together with several artistically twisted branches, and we would be ready to begin. Depending on the species to be photographed, Martin would position the snake either on the tree bough or on the floor of the studio in front of the rocks. That at least was the theory, and it worked well for the constrictors, a small anaconda and an emerald tree boa. This last snake was a joy to photograph, with its brilliant green body flecked with white and its habit of lying coiled over a branch like a garden hose, head in the centre and looking for all the world like a bunch of bananas from a distance. Although not venomous, the emerald tree boa has far larger fangs than its ground-living cousin, the boa constrictor. These fangs are necessary because the emerald, a tree-liver, feeds mainly on birds and needs the longer fangs to penetrate the feathers and hold onto its prey.

This easy run of events changed dramatically when Martin brought out some extra ground-livers. I watched in horror as he brought up cage after cage of venomous snake: a rattlesnake, two odd-coloured fers-de-lance, Delilah and, worst of all, a coral snake. Although it was less than two feet long, I knew a little about the red, black and yellow banded reptile that Martin placed gently on the studio table. The coral snake looks much less formidable than most poisonous snakes. Its front fangs are shorter than those of the bushmaster or even the rattlesnake, but it makes up for this shortcoming in its dental armoury by carrying an extremely potent neurotoxic venom. According to Pappy, the beast was fairly common on Lama Creek, where it was called the hymaralee and greatly respected. 'He make for bite you,' Pappy had told us, 'and even 'fore you reach to Land o' Canaan, you'se one dead man!' No-one at Lama would go within fifty feet of a hymaralee, and the same must be true for most other animals. So feared is this reptile that the scarlet king snake (*Lampropeltis doliata*) mimics both the coloration and the blunt-headed shape of the coral snake to save itself from the unwanted attentions of potential predators. And it was this snake that Martin wanted me to film!

To my surprise, the coral snake proved very tractable. It lay perfectly still throughout its portrait session. Later, I discovered that, although so poisonous, this species is extremely slow to anger. At that time, however, it was a very venomous reptile lying very close to me and I was absolutely petrified. It was fortunate I was using electronic flash to 'freeze' the picture; an available-light shot would have been terribly blurred, my hands were shaking so badly!

Delilah was the next subject. I was expecting trouble with this lady, but Martin defused the situation by using a large earthenware pot. He inverted it over Delilah as soon as she was placed in the studio and left her there in the darkness for a minute or two. Deprived of light, and in just the sort of small dark crevice she would naturally inhabit, the bushmaster relaxed and curled herself up comfortably. I focussed the camera just behind the front of the pot, and when Martin lifted it there was Delilah, beautifully posed for the photograph.

The next two customers were both fers-de-lance. By now it seemed almost natural to have venomous snakes six or seven feet from the front of the camera. They were not too bad really, I told myself happily, things were going fine. We had only been working an hour and already I had several species in the can. Bit of a doddle really. All you had to do was keep your head and treat them like any other animal.

But the coming of the fers-de-lance caused everything to fall rapidly to pieces. Both snakes zoomed off in different directions as soon as they made contact with the studio floor. I watched horror-struck as they fell from the set to the living room floor before Martin could bring his snake-stick to bear on either one of them. Then they were off. Martin went after the smaller fer-de-lance; it was moving more quickly than its compatriot and making unerringly for the slightly open front door. The larger snake was moving towards the innermost part of the room, and I was directly in its path!

By the time I thought of running, it was already too late; the snake was too close to risk it. There was nothing to do but 'freeze' and wait for the fer-de-lance to pass. I had placed my camera-case in front of me at the start of the filming session to act as a barrier; with my eye glued to the camera's viewfinder it was all too easy to move closer to the snakes than was wise. Here was a piece of luck! The case would surely deflect the path of the fer-de-lance, at present making for a point between my legs. But no – to my horror, the yellow pipe of poison was undeterred and slithered slowly up over the case until it was coiled on top. Here it stopped for a moment, testing the air with its darting, forked tongue. I was only a foot or so behind the case, well within Martin's golden distance, and quite rigid with fright. From its vantage point on the case the snake was on a level with my knees, and when it slowly raised itself and began inspecting my lower abdomen I almost fainted with fear. Surely it wouldn't, it couldn't bite me *there*! With a frantic effort of will, I forced my eyelids down over my staring pupils, blocking out the fer-de-lance's baleful scrutiny of my groin. If I hadn't, I'd have been tempted to run and that would have been disastrous. For what seemed an age I stood with my eyes screwed tight, sweat pimpling my face, imagining the snake swaying there below me and praying that it would go.

Then it happened: something struck me a blow between the legs! I screamed aloud and suddenly found myself a good seven feet from my original position, grasping my groin with both hands. My heart was racing and for an instant I thought I had been bitten, but one look at Martin's face told me a different story.

He was lying on the sofa, the fer-de-lance held in one hand and a snake-stick in the other, with a demonic grin on his face. He was giggling so hard that he could barely breathe. 'Oh man, oh man, oh man,' he wheezed, 'your face when I hit you with de snake-stick!' He mimicked my scream and leap. 'Man! You oughta be in de 'Lympics, you jump so hard!' He relapsed into another fit of giggles.

My first instinct was to hit Martin hard with the Amerindian club on the wall above his head. But he was still holding the snake and the last thing I wanted was to see that thing free again. Besides, I was gradually beginning to see the funny side myself, and it was not long before I too was giggling madly, though as much from relief as from pleasure.

An hour and three rum punches later, I was beginning to feel a little more composed. Liz and I were in the bar of the Park Hotel, poring over a map and trying to work out just which areas of Guyana her giant otter survey would touch upon and what techniques we would use. We had decided to solve the area problem in the time-honoured way: by choosing representative regions over as wide an area of the country as possible. That way, we would get a rough indication of the giant otter's survival status throughout Guyana. This information would mesh nicely with Nicole Duplaix's survey of the river wolf in Surinam. Together, these details would give a continuous area of more than 135,000 square miles in which both scientists and conservationists would have a fair idea of the health or otherwise of giant otter populations.

Nicole's survey technique had been to power along in an outboard-driven boat and to mark down any giant otter that she came upon, noting the number of animals in the group and, where possible, their age and sex. It seemed to have worked well in Surinam but an experience we had had while at Lama made us wonder if it would be as useful in Guyana.

We had been using the outboard to move quickly down Lama Creek to our new study site some way up the Mahaica River. About four miles from Lama, Liz suddenly pointed to an Ite palm on the left bank of the creek. Sitting in its branches were upwards of thirty orange-winged parrots, all feasting on the golfball-size fruits that hung in dense clusters from just below the crown of the palm tree. I cut the engine and we drifted silently forward to watch the parrots at their noisy feast. The current slowly pushed us beneath an overhanging branch on the right-hand side of the creek. In contrast to the temperature in mid-river, the air under the leafy canopy was beautifully cool. Our shade tree was in flower, too, and its yellow blossom, falling gently on the breeze like saffron snow, made our resting-place quite idyllic. There was even an absence of the usual mosquitoes and biting flies.

As we drifted, the sound of heavy breathing wafted quietly across the moca-moca. Then a deep, explosive 'hah!' reverberated through the greenery to our right. We looked immediately, but saw only a ripple of brown water striking against the moca-moca stems. But we knew that sound: it was unmistakeably a giant otter! Another 'hah!' and this time we were faster. There, not ten feet from the end of the boat – closer than we had ever been – was the head of a female giant otter, peering at us short-sightedly from beneath a lily leaf, her head moving rhythmically from side to side as she watched us through the branches. She seemed very excited and the heavy breathing we had heard was the air escaping from her red mouth in short gasps. An answering 'hah!' from a point further up the river had us spinning in our seats, and this time we were quick enough to see the distinctive white chest-bib as a second otter periscoped among the vegetation. It was Spotted Dick. Both he and Black Throat continued to keep us under observation for several minutes before slinking quietly away. The youngsters of the family were nowhere to be seen on this occasion.

The two otter parents had seemed very put out by these events, and little wonder. To the otters it must have seemed as if they were up against a pair of superb trackers who could somehow locate their presence even when they were tearing up the creek in an outboard-driven canoe. In fact, it had simply been an incredible piece of luck to cut the engine just where the otters had taken refuge in the creekside greenery. And even then, if they hadn't decided to scrutinise us we would never have known they were there. But this stroke of good fortune brought home to us the chancy nature of searching for otter groups from a motor-driven boat. Had we been on our otter survey and had we not stopped to watch the feeding parrots, we would have concluded that giant otter were absent from Lama Creek. Yet the truth was that Lama was prime otter habitat, and the otters simply scurried away into the underbrush whenever they heard the sound of an approaching engine. An outboard engine was like a police siren: it proclaimed our approach to the otters long before they became visible to us. Perhaps Nicole had used a more powerful engine on her Surinam survey, so that the engine's noise and the boat's appearance were almost simultaneous, allowing her to catch the giant otters before they had time to hide themselves. But for us it was obvious that an outboard survey would produce a terribly inaccurate picture of the river wolf's distribution in Guyana. It had to be paddle-power.

The camp sites in Dark Lane were another warning against a motor-driven survey. Dark Lane was a thin arm of Lama Creek which, with its overhanging branches and fallen trees, seemed impenetrable when viewed from a passing speedboat, but which hid a total of seven otter camp sites among its tangled network of lianas and branches. Only by using a native canoe had its defences been breached and this fact once again emphasised the advantages of a slow, careful survey of a small area over a widespread but superficial assessment of the giant otter's distribution.

Once we had settled on how we would look for the river wolf, we still had to decide where the survey should take place. The land of the coast was not very promising as almost ninety-five per cent of Guyana's one million souls live, farm and work within this narrow band of low-lying terrain. Behind this overpopulated ribbon of land our initial work had shown that a viable population of giant otter still survived in creeks such as Carabice, Maduni and Lama. How long they would remain viable in the face of increasing human expansion was anyone's guess, but firm action by the Guyanese government could no doubt secure 'no-go areas' – like Russel Lake – in which at least a portion of the coastal creek populations could survive and flourish. Now we had to look further afield.

Apart from the coastal area, much of the rest of Guyana lies within a range of mountains called the Guiana Massif. We chose the Upper Mazaruni Basin, centred on the small airstrip of Kamarang, as representative of this area. Martin and several other bush-men had recommended this as one of the best areas in which to see giant otter, before the High Commissioner had switched our attention to Lama. It would be interesting to see if they were right.

To the south of the Guiana Massif lies the Rupununi, a region of rolling, grass-covered savannah, liberally endowed with streams and rivers. This would be our second survey point, but one in which we expected little in the way of giant otters. The Rupununi savannahs are continuous with the grasslands of Brazil which lie just across the Ireng and Takatu Rivers. Several informants had told us that the river wolf had been shot in large numbers in the Guyanese savannahs and smuggled across the border into Brazil, where good prices were to be had for giant otter pelts, no questions asked. It was possible that the giant otter was already extinct in most parts of the Rupununi, and it was for this reason that we wanted to visit the region.

We looked at our map again: the coast, the west and the south of the country had been covered, what we needed now was a survey area somewhere in the centre of Guyana. We chose the stretch of the Potaro River above Kaieteur Falls. The sheer size of Kaieteur discourages most river traffic, and we hoped to find a goodly number of giant otter here. The area around Kaieteur Falls was also Guyana's only National Park. Though small by most standards – a mile-wide rectangle running for ten miles along the lower Potaro below the Falls – we hoped to have the park increased in size if we could show a substantial number of river wolf in the area.

With our itinerary decided, the final step was to arrange for permission to visit our three survey areas. Our original permits had expired by this time, so once again we began our interminable travels from ministry to ministry. Finally, almost unbelievably, all our papers were in order. As we travelled towards Timehri airport and our meeting with the plane that would take us to Kamarang, we were more eager than ever to be back in the bush. After six days in Georgetown the simple hardships of jungle life were as nothing compared to the endless sitting in humid offices waiting for one of the four separate permits needed to travel anywhere in the interior. In just a few hours we would be deep in the bush, as remote from civilisation as it is possible to be in this overcrowded world.

7 Tropical Klondike

(Keith)

The Upper Mazaruni Basin lies 200 miles due west of Georgetown, and more than 150 years in the past. Most of the country is
as virgin as when the first Europeans visited it. There are no roads
into the area and river traffic is non-existent because of the many
falls, cataracts and rapids that block the downriver stretches of
the Mazaruni. The only means of access is by aeroplane, with
landing facilities for suitable small planes at Kamarang and
Imbamadai. Few people travel to this remote spot; the only coast-
landers tempted to make the flight are 'pork-knockers', lured to
the Mazaruni by the promise of diamonds and gold.

The small indigenous population of some 3,500 souls belongs
mainly to the Akawaio tribe, Amerindians whose ancestors
crossed the land bridge between Russia and America at least forty
thousand years ago and were the first people to settle the then
deserted continents of North and South America. Scholars believe that the Akawaio now inhabiting the Mazaruni are descended from a group of Amerindians who, about 150 years ago,
were lured from their former homes by a holy man, an Arekuna
prophet named Awaikaipu who was convinced that he was 'God
on Earth'. He sent missionaries into the surrounding country,
telling the people to leave their lands and come to his camp
where, he said, a large crop of cassava could be harvested by
simply burying a single plant. This vision of a promised land
proved very powerful. There was a mass exodus of families, all
'going to see God' as the Akawaio put it. Many people perished en
route, and the remainder arrived in the Mazaruni tired and
destitute. Unfortunately 'God on Earth' was not able to deliver his
agricultural miracles and, to show their displeasure, one day in
1846 his disillusioned followers clubbed Awaikaipu to death.

Now most of the Akawaio have given up their old animistic
religion in favour of Christianity. Gone too are the reed skirts,
mouth plugs and blow-pipes, replaced with denims, sunglasses
and shotguns. But for all that, the Akawaio's basic way of life is not
all that different from their ancestors: they still hunt, fish and

plant cassava in the forest. A few have turned 'pork-knocker' and compete with the coastlanders for the best gold and diamond claims.

The tension of the gold-seekers pervades Kamarang, and we felt the electric atmosphere of the place as soon as we stepped from the plane. It was like being in a tropical Klondike, and the talk in the numerous rum-shops was of nothing but claims, shady deals, two-timing partners and big strikes made in places with romantic names like Kurupung, Aurora and Alligator Creek. The accommodation was as rough and ready as the rest of the town. While we sought out boats, engines and other essentials we stayed at a two-storey wooden building bearing the grand legend, 'Peter's Guest House: bed & breakfast'. The advertisement was perfectly correct, but apart from a bed and a breakfast, Pete, the genial host, could offer very little else in the way of comfort.

There were showers but, as they were worked by pumps and Pete was saving gasoline, actually getting water from the rose above your head meant a journey to ground level and a bargaining session with Pete, who always liked to know that your shower was really essential before he would condescend to turn on the power. Pete did have one shining virtue, however: he knew almost everyone in Kamarang and through his kind offices we were soon put in touch with the owners of several different boats and outboards. It was on the way to one such meeting that Pete pointed out a muscular Amerindian lying slouched against the side of a hut, his head in his hands and looking as if he was in the grip of a monumental hangover.

'See he?' Pete asked. 'Man, he one foolish Indian. Two days past he had 'nuff gold, 'nuff 'nuff!' Pete mimicked the carrying of the bullion. 'He work many many days in bush and come back wi' plenty-plenty ore. He could be in Georgetown now, feeding and drinking and eyeing de women.' Pete shook his head. 'But de fool, he go to nearest rum-shop, and he buy it! He buy it! Then he send to all his friends an' dey come an' he shut de rum shop door an' he an' dey stay in the shop 'til de rum all gone.' Pete shrugged. 'Now he got for go back to bush.'

We also had to get back into the rain forest, and for that we needed a boat. Although there are falls and rapids just a mile or two downriver from Kamarang, the Mazaruni and its tributaries – the Kako and Kamarang rivers – were all for the most part navigable in their upper reaches. But river transport was in great demand to ferry pork-knockers to and from their mines, and the going rate for even a small boat and engine was consequently extortionate. We hired the smallest craft we could find and set off

71

south-westward along the Kako River. On the river we saw pork-knockers working their claims, either panning for gold or using the more complicated and efficient 'sluice and tom'. On one occasion on the Mazaruni, we came across a large raft, supported by several large, forty-gallon oil-drums. As we watched, a brass-helmeted diver emerged from the water and climbed the rickety stairs onto the raft, carrying two containers. This was a diamond diver, laboriously collecting sackfuls of mud from the deepest pools of the Mazaruni. The mud is washed, sorted and passed through sieves, and the diamonds – if any – are picked out by hand. It is a dirty, dangerous job and fatalities among the divers are common.

The Kako River winds its tortuous path south-west, finally turning east as it reaches close to the base of Mount Roraima. It was in this region that Sir Walter Raleigh believed he would find the fabled, fabulously rich city of El Dorado, and also a race of men who had no necks! The country along the Kako River seemed very similar to Lama, at least from the boat. Yet it was daunting to know that the vegetation was not simply a strip of riverine forest as at Lama, but was instead a green sea that rolled endlessly across country until it broke against the high ramparts of the northern and western mountains, or met with Man's mechanical monsters to the east, where lumbermen were clearing hundreds of hectares of forest every day.

During our second day on the river we managed to find three giant otter camp sites – two fresh, and one which appeared to have been abandoned for several weeks, if not for months. Liz took scale samples from each of these sites with her six-inch grid and stored them for later analysis. She also found some strange-looking droppings on one of the sites, which Pete and another Kamarang resident later identified as ocelot (*Panthera pardalis*) faeces. This was the first time we had found the droppings of this cat species on an otter site, although we had discovered tapir dung – inch-long cylinders of matted plant remains – on one site at Lama. Why other animals bother to defecate on giant otter camp sites we do not know; possibly, a patch of bare earth is always utilised in this way, or perhaps the concentrated smell of the otters' 'loo' stimulates other species to defecate. Whatever the reason, otters were not the only animals to make use of the flat, bare latrines they had created.

Towards the middle of the afternoon, we stopped to make camp on the creekside. When we had first arrived in Guyana we had worked throughout the daylight hours and tried to begin camp construction only about an hour before sunset. The result was that we had not even time to make supper and sling our hammocks before darkness had descended and we had to fumble about as best we could with the aid of a small lantern. After that, we took our cue from the locals and ended our work day around 3.30 pm, when there was still a good three hours of light left. With this amount of time it was possible to clear the camp ground thoroughly with a cutlass (removing not only the vegetation but also any worries we might have harboured about lurking bush-masters and other poisonous snakes), and there was still time to collect brushwood, light a fire, set out the fishing lines and generally give the place as homely a feel as possible. There was even time to build a hammock-scaffold, a trellis arrangement made of saplings that allowed our hammocks to be slung at just the right length and height to give the most comfortable night's sleep. The fishing lines seldom failed, and although fish every night did get somewhat tiresome, it was always better than the dried lentils we carried in a large sack as emergency supplies.

After four days of this routine (and only two sightings of giant otter groups, each time very distant) we decided to ditch the outboard and to rely instead on paddle-power. Once again, we had found that the noisy approach of the outboard did little to secure good giant otter sightings; the silent approach was better. As soon as we arrived at what we judged to be likely otter habitat, we shipped the motor and moved quietly up the creek under our own steam.

We had almost passed them before we realised their presence. Deep in the vegetation, about ten or twelve feet from the river-side, were two giant otters, a mated pair by the way they lay so close together, and both fast asleep. Hardly moving our paddles, we drifted across to the other side of the creek and sat quietly, watching their afternoon snooze. They were lying stetched out against each other on the leaf-litter and it appeared that the animal on the right – a male – was dreaming, for his limbs jerked spasmodically like a dog chasing rabbits in its sleep. On one occasion the mosquitoes that were so annoying us must have become intolerable for the male also; he lifted his head and, eyes still firmly closed, shook his head violently to dislodge the biting insects. Throughout it all the female otter on his left maintained a peaceful, undisturbed slumber. From our knowledge of giant otter behaviour we assumed that this was a mid-afternoon siesta, which the species often takes after a long arduous fishing session. The catch must have been good that day, especially for the female. She was dead to the world, her belly quite distended with food, and halfway through her nap she developed hiccups! To see her huge belly heaving rhythmically up and down was a truly comic sight; each time she hiccupped, her body jerked against the male, spoiling his slumbers, but she continued sleeping like a baby.

The male could not get his rest under this continual abdominal barrage and it seemed that, like a lot of humans, once he had been disturbed, sleep eluded him. Even after his mate's stomach-spasm had subsided, he just could not get back to sleep. He fidgeted, scratched a little, rolled on his back, looking for all the world like a man trying to sleep on an uncomfortable bed. After a while he gave up; he lifted his head listlessly, yawned hugely, and began rubbing his neck across the flank of the female. Then he brought his mouth down on her shoulder and began to nibble the female's fur delicately, working up and down her shoulder in short, straight lines. His mate hardly stirred under this show of affection, but both Liz and I peered more closely through the gloom-obscured foliage – this was the first time in the trip that we had seen one giant otter groom another, though earlier observers had mentioned this activity. The behaviour, termed 'allogrooming' to distinguish it from 'autogrooming' (when an animal grooms itself), is very widespread throughout the animal kingdom. A whole range of animals groom, but none so assiduously as the primates, the order of mammal to which we ourselves belong. In the monkeys and apes, individuals who rank high in the social hierarchy are usually groomed most often and for longer periods of time, though some animals reverse this normal state of affairs; high-ranking members of South American spider monkey troops (*Ateles geoffroyi*) perform most of the grooming. If the normal pattern was followed in the case of the giant otter, then it would seem that the female ranks higher than the male: of five grooming sessions we observed, the female groomed the male once, and then only briefly. The main function of grooming among most animals is to promote group harmony and cohesion, and it is probable that this is also the reason for the behaviour in the giant otter.

I suspected another motive on the part of this particular male. It seemed to me that, having had his sleep disturbed, he now wanted to move on. The female, being so obviously replete and comfortable, wanted nothing more than to dream away the afternoon hours. Grooming was an ideal and rather crafty way for the male to waken his wife gently, bringing her back from the Land of Nod without incurring too much of her wrath! I smiled at the ploy: several days before I had used a similar ruse to rouse Liz from her siesta, in the hope that I could lie abed while she made the mid-afternoon coffee. It had not worked for me, and the male giant otter seemed to be having difficulty too. The female was quite happy to be groomed in her sleep!

Then, just to our right, a ringed kingfisher crashed into the water in pursuit of a fish. The noise of the ensuing splash achieved what the male's grooming had failed to do: the female sat bolt upright, instantly on guard, and both she and the male scanned the riverside suspiciously. Even with their slightly myopic vision, there was no way two alert giant otters could miss our boat or us. They rocked uneasily from side to side as they became aware of our presence, humming softly to one another for reassurance. There was nothing we could do but sit quietly, making no sudden movements, and wait. After several seconds' inspection both otters rose as if by a silent command. They looked undecided for a moment, then the female turned away and half-walked, half-ran into the forest. The male did not flee but backed away slowly, as if protecting her, until the vegetation had swallowed him completely. The instant he disappeared we made for their sleeping quarters, but our boat had no sooner touched the bankside than we heard two splashes further upstream. We hurriedly pushed off again, and as the boat slid away from the masking bankside foliage and into open water, we were just in time to see two ottery heads making upriver at high speed. We knew from past experience that it was impossible to catch giant otter by paddle-power if they were moving at top speed, and having no desire to frighten them even more with the outboard, we decided instead to check the site they had just vacated.

There was almost nothing there, just a few broken-down plant and bush stems near to where the pair had been lying, off to the right, and a small patch of spraint. Apart from that there was nothing, except for the rough track that the two had made while escaping. We followed this spoor and it led us to a small tributary of the main creek, into which the otters had plunged before swimming into the larger waterway. From the nervousness of the pair, the lack of offspring and the very rudimentary state of the otter site, Liz concluded that the two otters were a newly mated pair who were just beginning to set up their own territory. Perhaps they were establishing their home on another otter group's territory. That would explain why they had not been even half as confident as most other adults we met when paddling on the creeks.

After Liz had collected her samples and measured what there was of the site, we suddenly realised that there was not even two hours of daylight left – not enough time to set up camp properly or to make it back to the previous night's bivouac. Fortunately, we had passed an Amerindian village about noon, some four hours' paddling away. Faced with the alternative of roughing it in the bush, we decided to pay a return visit to the Amerindians and to sample their hospitality. The people of the interior are extremely friendly and we had no doubt of a warm welcome. We made the journey to the village by outboard (the two otters having swum off in the opposite direction) and arrived about an hour before sunset.

Amerindian hospitality proved to be overwhelming. Everyone, from the village chief to the smallest child, wanted to be introduced and to shake our hands. Who were we? Where were we from? How long had we been in Guyana? What were we doing here? We answered as best we could; filming for television was not readily understood until I realised that no-one in the village had seen a TV. When I changed my answer to filming for the cinema there was an appreciative 'uh-huh!' from several of the men who had travelled to Georgetown and experienced the delights of the silver screen. The questions were incessant and, after days of solitude in the bush, the presence of so many people standing so close and all jabbering away noisily was quite a trial. I began to feel that roughing it in the bush might have been preferable to being welcomed to death!

It was quite dark before things settled down enough for us to be shown to our hut, with an invitation to supper after we had made ourselves comfortable. Finally alone, we dipped into our kit bags to prepare our gear for bed. Liz went off to borrow some string for our mosquito nets, leaving me by myself in the hut. My first task was to light the small butane gas-lantern we always carried. I put a match to the mantle and the lantern burst into life. I looked up and the first thing I spied were the luscious breasts of Miss Sian Adey-Jones, posing topless on page three of *The Sun*! I could not believe it! Had the jungle finally got to me? Was I hallucinating? I had to be! The only sun in Guyana was the one that burnt your skin and parched your throat. I looked around the hut and the feeling of unreality increased. Not only did Miss Adey-Jones grace the wall, there were four other naked lovelies fastened to the wooden posts of the building. The night wind caught one of these, lifting it to reveal, in the flickering lantern light, the exhortation 'Visit Photomarkets for More Camera Bargains NOW!' That was the final straw. I knew I was going off my head. Where was Liz? She would help, I knew.

I started for the door just as Liz re-entered with her borrowed string.

'Do you see them?' I asked in a panic-stricken whisper. 'Can you see them too?'

The look on Liz's face reassured me at once. She was slack-jawed with astonishment and it was obvious that, unlikely as it seemed, there really were topless pin-ups from an English daily newspaper on the walls of an Amerindian hut deep in the South American jungle. The question was: how the hell did they get here?

At our meal with the Amerindian elders later that night over a pot of stewed tortoise, we talked about the giant otter. There were still many about on the rivers, according to our Amerindian friends. They were rarely hunted in the Upper Mazaruni Basin and most of the otters that were killed were shot by the coast-landers. The pelts apparently made very good slippers, which were much in demand by some sections of Georgetown society. Occasionally an Akawaio would kill a giant otter either by shot-gunning it on the river or by spearing the animal while it slept on the bank at night, but only when the fishing was very poor. As at Lama, we heard the same tale of otter packs storing their fish on the camp sites and sharing the catch equally between pack members so that, although neither we nor any other trained zoologist had seen this behaviour, we kept an open mind on the subject. After all, the Akawaio had had much more time to observe the giant otter than any white person.

Eventually I summoned up the courage to turn the conversation towards the page three pin-ups. What, I asked, were they doing in the village?

There was an embarrassed silence; then the chief, a small, wizened man with skin like parchment, explained. An expedition, he said, had come from England, from 'de big garden where dey all queue'.

'You mean Kew Gardens?' I asked.

'Right, right,' said the old man, sucking on a peccary bone and oblivious of his blunder, 'and dese plant-men, dey cover all de t'ings with paper, to stop dem break-up.' He grinned, and his old eyes, red in the reflected glow of the cooking fires, danced with devilment. 'Some paper, some paper have de white woman picture – very, very good picture.' The rest of the elders nodded sagely in agreement, and appreciative comments could be heard from the young men ranged all round. 'So,' the chief continued, 'some of our people, dey keep dem for decorate de house.'

It turned out that this village did not really have a good selection. The Kew expedition had been bound for Roraima, the huge, flat-topped plateau of *Lost World* fame, and the closer one got to Roraima, the better the display of 'white woman pictures'. The village where the expedition unpacked had, according to our envious informant, huts whose walls were entirely covered with these topless teasers. They were very highly regarded and there was even a flourishing trade in second-hand pin-ups! It was almost too much to bear and I had a hard time preventing myself from laughing as several of the younger members of the village discussed the going rate for some of the ladies. Apparently one Miss Adey-Jones was worth ten shotgun cartridges or two medium-sized tortoises!

Four hours later, as I leaned out of my hammock to switch off the gas-lantern, the last thing I saw was a semi-naked Miss Adey-Jones, kneeling suggestively on a sheepskin rug with a provocative 'come hither' expression on her pouting face. It made a pleasant change from bats and cockroaches.

The following morning we were on the river again. It was a blistering day and to our hot, sweating bodies the creek water looked terribly inviting. It was pure purgatory; I felt like King Tantalus in Greek mythology, seeing so much cooling water so close and knowing I could never enjoy its invigorating balm. It would have been foolish to succumb to its refreshing temptation; not only were there piranha and anaconda in the river, but we had heard tales that the dreaded candiru also inhabited this area. The candiru is a tiny species of fish which is attracted to the urethral openings of both men and women, apparently drawn by the scent of urine. Once it has entered, for example, a man's penis, the backward-pointing spines on the head make it impossible to remove the fish without hospitalisation. The fish, of course, dies and rots in the urethra; the pain is said to be exquisite.

We saw no more giant otter that morning, and in the afternoon we pulled into the bank to prepare our camp for the night. We had finished clearing the site and setting up our hammock scaffold, and were preparing a meal when Liz claimed that she heard a call: the snort of an otter. I was sceptical, but Liz has very good ears so we dropped everything and made for the boat. As we climbed aboard I remembered the cameras still in their case, next to the hammocks. I should have gone back for them, but it was late, we had had a rough day and besides, I told myself, the light was not much good. I will regret that piece of self-delusion until my dying day.

I heard the otters after about a minute's paddling. They could hardly be missed, for an incessant barrage of squeals, snorts, barks and grunts had begun to ring down the creek. We turned a bend and found that we had stumbled upon a dramatic and probably unique piece of giant otter behaviour. About thirty yards away, a family group of six was ranged in a rough semi-circle around the remains of a drowned tree close to the bank. Hanging in its branches was the biggest anaconda I had yet seen in Guyana. It must have been all of fifteen feet long and as thick as my thigh. We immediately tied up against an old tree stump, entranced by the wild spectacle before us.

The giant otters were mobbing the anaconda, intimidating the snake with violent growling rushes towards its position in the upper branches. The adults and the two sub-adults in the group made most of the running; the two small cubs, though vociferous in their vocal assaults, milled around in the water behind their older relatives and stayed well away from the anaconda. On one occasion, the big male became so bold that he actually pulled himself half out of the water onto the main trunk of the sunken tree. He soon thought better of his bravado and leaped back into the creek when the anaconda made an open-mouthed strike in his direction. But the snake could not follow up its attack; by that time another otter was already attacking.

What had prompted the otters' assault we did not know. It may be that the giant otter always mobs these huge reptiles whenever chance places one in its path. Faced with such a deadly potential predator, attack may actually be the best form of defence. Or perhaps the anaconda had tried to secure one of the young cubs for its next meal, forcing the otter group to protect its own. Whatever the reason, the mobbing behaviour was certainly having the desired effect. The anaconda seemed totally confused by the constant, random attacks from every quarter, and if it had ever harboured any malevolent intentions towards the cubs, it was evident that now its only desire was to effect a safe and speedy retreat. After a particularly violent sally from the adult female, it snaked further along a branch close to the bank, watching its attackers' moves with flat, reptilian eyes. This action proved its salvation, for the weight of the snake caused the branch to bend until it was almost touching the creekside. With a speed I found astonishing in so large an animal, the anaconda slithered quickly along the branch, dropped to the ground beneath and was rapidly lost in the creekside underbrush.

The otters were as surprised by this sudden disappearance as we were. They jostled about in the water and seemed to feel quite peeved that their enemy had left the battleground so soon. The otters were still highly excited, but none seemed willing to pursue the anaconda on land. The most they did was to wander clumsily about in the shallows, the adult male and female touching noses several times as they passed each other, and still calling out occasionally. We remained very quiet during this period: there was no knowing what the giant otters might do if they spotted us while they were in such a high state of arousal. I was especially worried as, along with the cameras, the cutlass was also lying uselessly back at camp so we had only our survival knives to pit against the sextet's needle-sharp teeth if it came to an attack.

Fortunately, the giant otters were so preoccupied in celebrating their recent victory that they were completely unaware of our presence. They stayed around the sunken tree for several minutes, as if hoping that the anaconda would re-appear. Then they turned and, still chuntering amongst themselves, headed off upriver. We began paddling back to the camp site, overjoyed by our good fortune. The following morning we had to return to Kamarang to prepare for our departure to Georgetown en route to Rupununi. They might not have known it, but the giant otter group fast disappearing upstream had given us the best farewell present we could have asked for.

8 Niblet

(Liz)

We had just finished a good-bye drink with Pete, the hotel-keeper, and some of his friends and were packing our things to leave, when Pete suddenly yelled from downstairs that someone wanted to see us. A stocky Akawaio Amerindian was sitting on the front steps, holding a Banks Beer carton in his lap.

'I think he wants to sell you something,' said Pete.

'Beer – great! What miracles did he have to perform to get that here? Haven't had one in weeks.'

Pete grinned at Keith. 'Sorry to disappoint you, Doc, but I don't think it be beer.'

The man remained silent. He wore one of those neutral, self-contained faces that come so naturally to South American Indians, the sort of deadpan immobility most businessmen strive to cultivate. Then he opened the box and pulled out an otter by the scruff of its neck. 'I sell you this for one hundred Guyana dollars,' he announced.

It was a *Lutra enudris*, less than half a metre from nose to tip of tail and probably no more than three months old. It seemed intent on running away but the Indian held it firmly by a collar.

'How did you get him,' I asked quickly, 'and where?'

'You interested for buy?' the man repeated, ignoring my questions.

Keith attacked. 'Do you realise that you have broken the law? These water dogs are very, very scarce and no-one is allowed to kill them.' Actually, this was not true at all, and Keith knew it, but it was the best approach to take. The little otters were scarce, yes, but sadly enough there was no law to protect them in Guyana. Only the giant otter had that privilege.

The face remained implacable. 'Sixty dollars. Wid de collar.'

We knew we had the advantage, as we were the only customers for miles around; there would be no hope of a sale to any of the locals. Not that the little otter lacked appeal as a pet, quite the contrary, but a villager in these parts would be hard pressed to find enough fish or meat every day over and above his own family

81

requirements to satisfy an otter's high-protein appetite. A meal takes less than an hour to pass through an otter's gut and to cope with this extremely rapid digestion the animal has to eat about a quarter of its body-weight in fish every day. I remember Pappy telling us that bush-dwelling Amerindians are very skilled at catching young otters but they do so on the spur of the moment, not looking ahead to the problems this kind of pet will bring. In nine out of ten cases, the parent or parents were killed in order to procure the cub so even if it was returned to the place where it was captured, the chance of successful rehabilitation without parental help would be very slim. An easier option would be to dump it into any old river, whose resident otters might not tolerate an intruder, chasing it out or perhaps even killing it.

'What would you do with him if we don't take him?' I queried.

The man shrugged. 'Drown he, mebbe, or throw he back in de river. Me can't feed he no more. He eat too much.'

I believed him because it was the only reasonable thing for him to do. There was no way we could desert the little orphan. To do so would have been to sign its death warrant.

Keith spoke. 'We'll pay you a third of the asking price, twenty dollars and no more. I don't see why you should make a big profit out of this. This water dog will be more trouble than enough to keep, so we aren't really all that struck on having the beast.' Keith paused dramatically. 'Take it or leave it.'

Pete grinned. He must have wondered where we had learned to drive such hard bargains. The Amerindian gave way. He lowered his eyes and nodded, pushing the furred creature towards me with one hand and holding out the other for the money. A measly profit was better than no profit; it would at least pay for the food he had given the worthless animal.

Before our taciturn salesman left, I managed to extract some information from him. Yes, he had killed the parents on the creek with his shotgun and had caught both their cubs. One died in its first week of captivity. And no wonder – he had fed them on little else but roti (a kind of oily pastry) and white rice! I was surprised that this one had managed to survive so long, kept on a diet like that for five whole weeks. He was far from a picture of health: he had a toast-rack of a rib-cage and his fur looked sadly neglected. Even so, he made to follow the Amerindian, showing that some sort of bond had formed between them. Better the devil you know than the one you don't, I mused ruefully.

'C'mon, little otter,' I said, picking him up, 'let's go find some food for you.' But he objected immediately and slid like a piece of furred liver from my grasp. He made for a clump of bushes at the

82

Russel Lake, east canal

A familiar sight, the broad-fronted caiman

Katina with his trophy, a dea
caiman. The missing left forefoo
is the work of piranhas

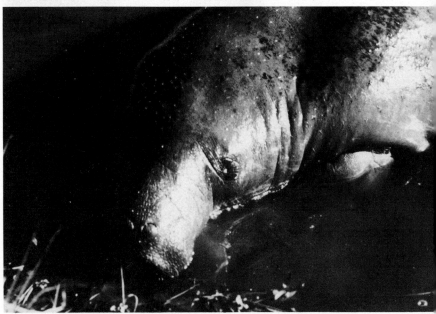

The manatee, a former lake-dweller, now endangered

Lama Creek

t home among the forest branches: the ocelot

Three-toed sloth

Marking site on east dyke

Freshly deposited scent

Close-up of latrine, showing scales and bones

The Streaky group periscoping

Mister and Missus

A streak of silver – the first sign of otters ahead

Resting in between hunting bouts

Hunting in the shallows

Above and left: Feeding in the shallows

Below: Landing a fish

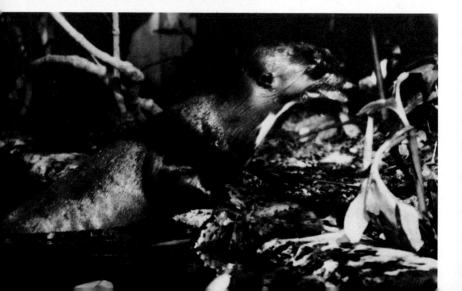

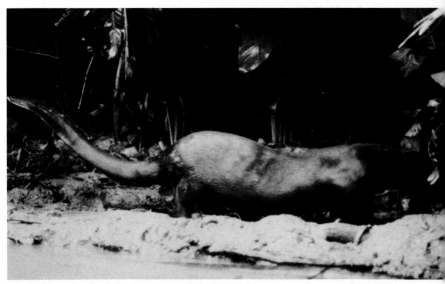

a, b & c: Excavating a signal spot

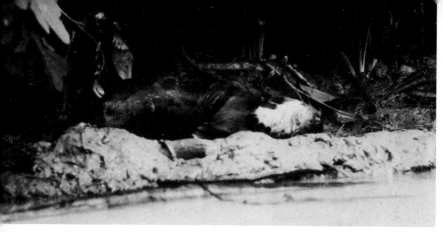

Sleeping

Grooming

Bluff-charging the canoe

Giant otter skins and seller in Georgetown

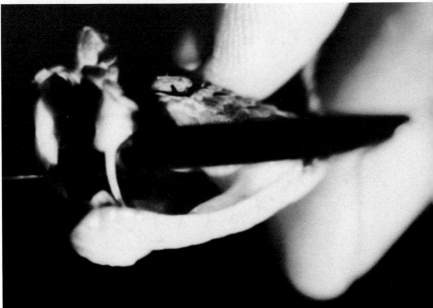

The fangs of a *fer de lance*

Niblet – his first night in the Park Hotel

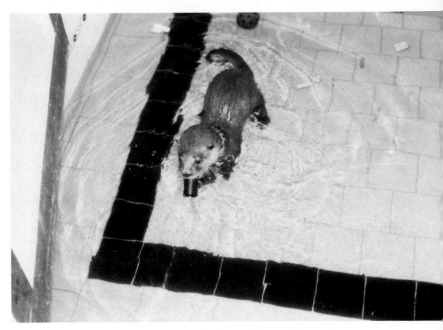

Nibs at play in the rest-house shower

Niblet gorging on a meal of patois

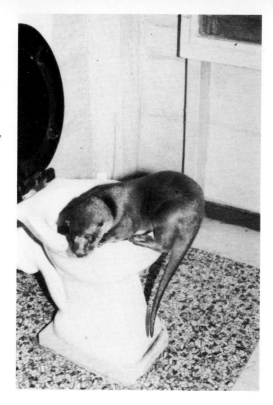

A favourite oasis for a thirsty otter

A weekly washday hindrance

Waiting to be chased out of the fire bucket

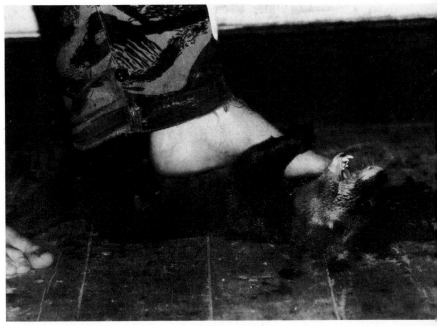

Wobble-belly

The metronome

Feet are never safe with an otter around

Saved by Keith after a clos
call with the weeds

The day of Nibs' release

corner of the house and refused to respond to our coaxing calls. If we tried force, we feared he might run further afield and we might lose him for good, so we decided to entice him out with food. Pete offered to send some boys to the river to catch some fish with his cast net. As we waited on hands and knees for the supplies to arrive, I suddenly remembered our planned trip to Rupununi. The Amerindian's unexpected visit and our excitement at saving the baby otter had made us forget about it completely. We had a plane to catch – and fast! The next scheduled flight to Georgetown was one week from now, and we could never be sure that it would not be postponed because of bad weather. From Georgetown we would board a second plane, bound for Rupununi and our survey. But that was before our new arrival: it was beginning to dawn on both of us that our newly acquired protégé was in no fit state to travel so far or to undergo the unpredictable hardships of expedition life. We agreed to catch the plane but to postpone the Rupununi trip and stay in Georgetown and Lama until the infant had recovered his health.

The boys returned with two pounds of finger-length sun fish. I reasoned that since the cub had been caught fishing with his parents, he was probably about three months old (the weaning age for the European otter, another *Lutra* species) and should be quite capable of eating fish. He had certainly managed to deal with the solids the Amerindian had given him, soft though they were. Keith flung a whole fish on the ground in front of him. He sniffed it for a minute or two and then snatched it between his jaws. After walking a few paces, he lay down in typical otter fashion, his high-arched rump making him look like a three-dimensional question mark. He crushed the fish's head with his large, multi-pointed back teeth, all the while holding the tail firmly in position with his forepaws. When the fish was half-eaten, the grip on its tail was released and he snapped his head backwards so that his skull and jaws were almost at right-angles to his body. This was our first demonstration of his yoga-like ability to bend body and limbs in all directions. I thought he was going to choke but he now had the fish half where he wanted it, on the mortar and pestles of his molars. Every now and then, the corner of his lips moved back off the formidable arsenal, giving him a rather evil look.

A pound of fish later, our little water dog walked away from the remains of his meal and started up the outside stairs to our room. We followed him cautiously, catching up with him as he loped towards the sink. He sniffed the drainage pipe thoroughly and moved on to the next corner of the room. The investigation was carried out in silence. Every piece of furniture was given the smell test and, having completed the circuit, he gave a last sniff to the air in general and trotted back over to the sink. We stood back and wondered if he was going to repeat the whole procedure over again but no, something under the sink had caught his attention. He suddenly swung round and faced us, all four legs placed well apart, webs splayed out, tail raised off the floor. He looked away for a moment and then I realised what was coming. A little pool of water spread slowly round his nether regions, followed by a small black blob. He moved away without so much as a backward glance, quite indifferent to the fact that he had done his business in the most inconvenient spot in the room. Little did we realise then that he was to preserve this partiality towards sinks as sprainting spots. The little otter had us infatuated fair and square and we were blind to all future problems.

We called him Niblet in honour of his preferred pastime: nibbling our toes and fingers. We regarded Niblet as 'he', but in fact we weren't sure we had chosen correctly for several more days. As a name, 'Niblet' had the advantage of being bisexual – add a 't' and an 'e' and 'Niblet' became the very feminine 'Niblette'. He gave us our first taste, or rather feel, of his needle-like canines as soon as he had relieved himself. Rolling onto his back, he grabbed Keith's foot in his stumpy forepaws, leaving his back legs to flop about in a state of fluid relaxation completely disengaged from the front-line action. Occasionally, one would be drawn up to help grasp a stubborn big toe or a tricky piece of ankle, but otherwise they were left to laze. Their owner was on cloud nine, a picture of mischievous contentment. I was extremely relieved he had taken to us so quickly. He had looked so pathetic and lifeless when the Indian had pulled him out of the box and I had worried that the trauma of a new foster family might finish him off. Some otters become very attached to their owners and, if separated, will pine deeply for them. The heartrending result is an unhappy, neurotic animal.

Keith was finding Niblet a handful, yelling in pain every time he made a sideways lunge. In this position, his canines and molars operated more effectively and Keith had to shift his toes back quickly into the less painful frontal grip. There, only the small incisors came into play and as they lacked points and a strong grip, they were actually quite bearable.

We barely made Kamarang airfield in time for the Georgetown flight. With so little time we were unable to make a proper carrying container for Nibs, and we had to make do with the same beer-box the Amerindian had brought him in. The thought of forcing him back into it broke my heart, but Niblet surprised us all by readily leaping in. Over the weeks with the Akawaio hunter Niblet must have come to regard the cardboard container as home. But as we boarded the plane, we were still quite apprehensive. We could not be sure how he would respond to new situations; if he escaped in the plane we would be in trouble. There was nothing in the airport regulations against having otters running free in the passenger cabin, but our little friend had already proved the nosey type and I had no wish to distract the pilot's attention.

As it turned out, no-one even bothered to check the box, such was the general acceptance of beery habits. Niblet remained conveniently quiet during boarding and take-off, but this happy state of affairs soon changed: 2000 feet above the rain forests of the Mazaruni, a burst of 'wheep-wheep' suddenly cut across the low cabin hum. I could see other passengers tensing at the unidentified noise. Was something wrong with the plane? Only a couple of weeks previously, a similar plane had crashed in a remote part of Guyana, killing all the occupants. It took us a full ten seconds to realise that the staccato sound had come from Niblet's box. Just to make sure we knew, he had another go, but this time the tirade went on and on, riddling our nerves. I peered through one of the holes on either side of the box and met Niblet's big eyes staring straight at me. My heart went out to him, he looked so small and helpless in there. By all rights he should have been swimming in some nice fish-laden creek with his parents and his brother. Instead, here he was cooped up in a dark coffin with every minute taking him further and further away from his home waters.

I untied the box and lifted the flap. He flung himself between our legs and I was glad then that Keith had insisted we tie a nylon cord to his collar. But it was over five feet long and left Niblet well within the range of the man's legs in front. He jerked his feet up as if he had stepped on a nail and let loose a loud Guyanese expletive. I hauled at the lead and brought the cause of the curse to heel. He refused to be picked up to play bite-the-finger on our laps so there was nothing I could do but let him pot-hole under our seats. This new subterranean environment he seemed to find very interesting though I shudder to think what condition we left those seats in! The nerve-paring 'wheep-wheeps' had stopped and, apart from the coloured chap in front whose ankles were still recovering from the shock of being kissed by an otter, everyone else sank back into the stupor of flight. Niblet in his turn ignored us all and got on with the business of taming our seats from below.

85

The thought passed through my mind that he might be destroying our parachutes or life-jackets under there, but I was more concerned at the moment with keeping Niblet happy and silent than with survival strategies. My one fear was if and when the urge came for him to spraint. If that happened it would not matter if he was noisy or not: where otters are concerned, silence is no substitute for smell. But I need not have worried. Nibs had no desire to stake out his territory on a plane and he kept his mind on the task he had set himself, coming up only for air and a reassuring tickle.

At last, the 'fasten-your-seatbelt' sign came on for the landing and we dragged him out of his Aladdin's cave and put him in his box. It took a few bribes with our fingers as the prize, but we eventually got him in, tail apart. We had to let that stick out through one of the side holes. But fearing an escape attempt at any moment, I did not really relax until we were in the taxi on the way to Georgetown.

We booked in at the Park Hotel for a couple of nights before journeying on to Lama. We had stayed there before and knew which areas of the hotel the manager frequented and therefore which to try to avoid now. Keith signed the register while I hid in a shadowy corner with our forbidden baggage. Formalities completed, we hurriedly took our luggage – and beer-box – across the courtyard towards our room. But fate was in the mood for a little caprice in the shape of Mr Singh, the under-manager. He passed us in the foyer, just as Niblet decided he had had enough of his claustrophobic coffin. With a superhuman effort, our pet pushed his head through the side hole to see whom we had bade 'good evening' to. Mr Singh stared at the furry apparition for a moment and then asked politely 'Is that a mongoose?'

'No,' I said, trying to stuff the head back through the hole though with little success. Flattery was the order of the day. 'But it looks like one, doesn't it? They belong to the same family, you see,' I buttered. 'You're the first person to notice the resemblance.'

'Got to take him over to the zoo this evening,' said Keith by way of reassurance. 'They want a mate for the female they've got there.' Keith is one of those people who can lie through their back teeth with a straight face, and I was more than grateful for this talent now as there was no doubt in Mr Singh's mind that this was exactly what we were going to do. After an endless monologue on how he used to visit Georgetown Zoo every day as a boy, we finally managed to part company and escape to our room. I thanked God he could not tell the difference between grins of interest and grimaces of fear!

With the door locked behind us, we breathed a little easier and turned our attention to extricating Niblet from the box. His head was still sticking out through the hole and try as he might to pull it back in, he could not. His loosely worn skin kept furrowing into folds round his head and made him look like a miniature walrus. But on the fifth try he made it, just as Keith was about to cut the box. We let him explore the room in peace while we showered and changed our clothes. No corner was neglected, no furniture leg left unsniffed. Only after he had vetted every square inch of the room did he come into the adjoining bathroom and spraint – right under the sink as before. As Nibs sprainted frequently, the swiftly growing pile of ordure created something of a bottleneck at the entrance and left us little room to negotiate our way to the human loo. But the siting of his action centre beneath the sink was immutable. We tried transferring his spraint to a more convenient spot but Nibs would not be moved, and we eventually gave in and learned to accept it.

In the weeks that followed we had to do a great deal of accepting. Compromise is not something otters know a great deal about and more often than not it was us, the poor owners, who were controlled. On one occasion at Lama I was foolish enough to think I had found the secret of calling Niblet to heel. He ran towards my snapping fingers and I felt a surge of power at this show of obedience, the beginning, I thought, of a new and easier relationship. But it turned out to be an illusion: the apparent transformation to dog-like obedience had less to do with my ability to command than with the faint but detectable smell of fish on my fingers!

Next morning we had to rush off to the zoo to see what progress had been made in our absence on the giant otter enclosure. We wanted to get to Lama with Nibs as quickly as possible and there was little time to oversee the tank construction and buy supplies. It was out of the question to take Nibs around with us in his box so we fed him on a hearty breakfast of raw mince and left him in the hotel room with a 'Please don't disturb' notice pinned up outside on the door. We did not want to risk an escape.

When we returned to the hotel the receptionist called us over and informed us rather stuffily that Mr Ali, the manager, wanted to see us. Keith asked why, and the girl smiled cryptically. 'I don't really know,' she said. 'De maid see somet'ing.' I suddenly realised just what it was that Mr Ali might want to discuss, and my worst fears were confirmed on entering our room. It was not as we had left it, not by a long shot. The waste-paper basket had been overturned and its contents were strewn all over the floor –

orange peelings, mango seeds, newspaper and shredded tissues. The two bathroom towels had vanished from the hand-rail and lay in a sodden embrace on the carpet between the two beds. Beads of mince had hitched a ride on them, though some bits had fallen by the wayside, leaving a spotty trail of red. With rising horror, Keith pushed open the bathroom door and was almost thrown back by an overpowering blast of ammonia. It came from a shallow pool that circled the loo and completely masked the smell of the spraint beneath the sink. We had actually caught a whiff of it from the corridor but I, poor innocent, had thought someone had dropped their smelling salts. Otters' urine is particularly pungent on account of their high-protein diet, but it was the first time I had experienced it in such a confined space. I saw and smelt the chaos of the room and understood Mr Ali's urgent invitation.

'Oh my God,' said Keith, 'this spells "Out" for us.'

'Ah well,' I said, 'we were only going to spend another night anyway, though I really don't fancy our chat with Mr Ali. By the way, where is Niblet?'

We could not see him anywhere and I was beginning to think that he had escaped through the door when the maid had opened it. But then Keith spotted half a tail sticking out from under a hammock in one of our open suitcases. The folds of cloth rose and fell evenly in innocent sleep. As we watched, the tail twitched slightly and a back leg poked through the cloth and 'ran' a few paces. I was reminded of what one owner had said about the limb movements of her European otter cubs during sleep. They were, she felt, 'a re-enactment in their dreams of the frolics of their waking hours'. We left Niblet to enjoy the action replay of the havoc he had wrought and got on with the job of cleaning up.

Mr Ali was more understanding than he should have been. His mild mien showed no trace of anger and he spoke in his usual solemn voice. He had been making a check of all the vacant rooms that morning when he and the maid caught a whiff of ammonia in the corridor. They traced it to our room and on opening the door not only found everything in a smelly mess but a 'furry brown animal' was running round the room. What was it, he asked.

'An otter,' we replied in unison.

'We're going to give it to the zoo,' I lied rather irrelevantly and without a tenth of Keith's expertise.

Didn't we know, said Mr Ali, that the Park Hotel did not allow pets of any kind into its guest rooms?

No, we were really very sorry but we had not realised this (though we had rather suspected it). 'We don't expect the maids to clean up anything,' I assured him. 'We've already done it. Believe it or not, otters are really very clean animals. They do their business in one spot and keep their fur spotless. It's just that we were in a hurry this morning and had to leave before cleaning up the spraint and wee from the night before.'

Mr Ali nodded calmly and then surprised us by asking all sorts of questions about Niblet and his kind and about my giant otter study. We ended up having a very pleasant conversation, a far cry from the wrath and embarrassment I had dreaded. But the episode had taught me a lesson: never expect conformity from an otter – they are creatures of independent spirit and, unlike dogs, have not the slightest desire to please their human owners.

We were so relieved not to be kicked out of the hotel that we indulged Niblet that evening, though keeping an eye on the mess we made. We tickled him all over and let him play with our fingers and toes, his little mouth opening with 'ah-ahs' of delight. We hauled him up onto the beds and played hide-and-seek among the sheets and pillows. Twice when he trapped himself in a pillow case and could not find his way out, he wheep-wheeped plaintively for help and we had to drag him out by his back legs. In the course of the evening, a new game was introduced called 'butchers'. I excavated it from my childhood days when my mother would pretend to be a butcher and saw off pieces of my anatomy for sale, at the same time giving the customers a running commentary. 'A slice of leg, madam? Yes, certainly!' And the leg would be sliced off with a vigorous movement of her hand. 'A pound of bum-bum? Hmm, I think we can manage that.' Toes, head, chest and arms all got the same treatment and when she ran out of anatomy she would start all over again. Niblet loved this one; it meant lots of close body-contact which he seemed to need and he had us doing it over and over again. The guests on either side of us must have thought we were either crazy or quaintly perverted!

When he at last began to look tired, I took him in my arms and he automatically held my thumb between his front teeth, clasping my wrist with his forepaws, while I swayed him from side to side in a gentle lullaby. His big brown eyes remained wide open but he let his body go limp. He was, after all, still a baby with the same need for contact, comfort and security as a human infant. As he got older, I found that Nibs began to predict the movements. He would throw his head first to one side and then to the other, in time to my to-ings and fro-ings. It seemed that the rhythmic rocking was very pleasing to him, and the game became so popular that we even gave it a name, the metronome.

But the metronome game could not last forever. We were starving and had to go out for something to eat. I reckoned he was ready for bed after all the energetic fun and games but Niblet thought otherwise. Tired he may have been, but he had no wish to go to sleep right away and he certainly did not want us to leave. The wheep-wheeps started when I put him down in a corner of the room and followed Keith to the door. We just managed to close it before he came bounding over in protest but the calls that followed were so heart-rending and so loud that we knew we could not ignore them. Keith went back in and dangled one of his shoes from the back of a chair to see if he could distract him. But it was no use and we ended up turning on the air-conditioning unit to drown the insistent tirade.

Three-quarters of an hour later we came into a silent room. There was no enthusiastic greeting and we found Nibs lying on his back under the bedside table, juggling with a ball-point pen in a rather desultory fashion. He gave us an off-hand glance and went on playing by himself as if to say, 'I don't really need you, you know.' But the coolness vanished when his supper appeared and he bounded over to the plate before Keith could make the 'tch-tch' call, a signal he was already learning to associate with food. He demolished ten one-ounce fish in short order and, belly full, gave us a forgiving nibble before disappearing under the bed for the night.

By mid-afternoon the next day we were ready to leave for Lama. Laden with our basic food supplies of split peas, rice, flour, greens and dried shrimp – and a beer-box of otter – we caught an east coast taxi at Starbroek market near the wharf. These taxis are geriatric Morris Oxfords that have been coaxed reluctantly out of retirement by the Guyanese, and which are regularly overloaded with as many passengers as can be squeezed in. They were, one driver told us, called the 'English Jackass' in praise of their stalwart endurance. What he omitted to tell us was that the shortage of spare parts in Guyana also made them dangerous to drive in. But as £1.50 was all we paid for the same trip that a 'respectable' taxi would charge £15 for, we decided to swallow our fears and take the risk. With Nibs to cope with that afternoon, we were not looking forward to the hour's ride along the coast. The journey would take us east of Georgetown to Mahaica where we would have to catch another taxi to Cane Grove, a rice-growing village about a mile from the north-eastern corner of Russel Lake. We had done it several times before but not with a baby otter. It took us three-quarters of an hour to grab a sufficiently empty car, a grey contraption whose battered contours bore a faint resemblance to a Morris Oxford. We sat in the front with the driver, Keith on the outside with Niblet's box on his lap. I pushed our supplies into the gutted dashboard below a psychedelic sticker that read 'I'm sexy – touch me', and waited for the car to fill up. We set off with a respectable half a dozen on board but picked up another four en route, which resulted in two layers of humanity in the back and one and a half in front. A little Indian girl was poised precariously between me and the driver and I could only marvel at the man's ability to change gears through such a tangle of knees.

Niblet had been remarkably quiet all along. He contented himself with the odd thrust of his head through the side holes of his box for fresh air and a tickle under the chin. Perhaps it was the loud chatter and shrieks of West Indian laughter from the back-seat occupants that caused him to change mood or perhaps it was the intense heat of the afternoon. Whatever it was, Nibs started to create a mighty rumpus about halfway through the journey. He wheep-wheeped in ear-splitting volleys and ripped through one of the side holes of his box before Keith could shout 'Niblet!' The little girl adhering to my right side shrieked in fright as Nibs lunged across me and clawed at her skirt. Keith and I both grabbed at Nibs' collar and tried to reassure the girl and her relatives behind that 'the "beef" doesn't bite'. Now was certainly not the time to draw fine distinctions between an aggressive and a playful bite. The important thing was that they both boiled down to a certain amount of pain so at the moment prevention was better than cure. We kept our little ball of fur on a tight rein but he still managed to climb up round my neck and put the poor girl into hysterics every time he brushed past her hair.

Still, Nibs was coping very well with what to him must have been an extraordinary situation. But he was beginning to get hot and I was worried when he started to pant and drool saliva. I knew that otters can quickly overheat and die if no steps are taken to cool them so I asked the driver to stop at the next standpipe. It is amazing what water can do. After slurping his way through three calabashes of the stuff and gargling pleasurably with the jet, Niblet was a completely new otter. Small wonder he was loath to get back in the car, but this time I managed to keep him cool with a canful of water I had collected. By the time we reached Mahaica, Keith and I were as soaked as Niblet and felt a lot less comfortable than he did.

The second taxi-drive was even worse. The road to Cane Grove – if it could be called a road – was a nightmare. Huge pits and ridges stretched out in acne'd perspective and created havoc with what was left of the car's suspension. My own female shock absorbers ached from the constant buffeting and several times Niblet was nearly thrown out of the window. Suddenly, there was a terrific bang and the car slewed across the road, brakes squealing. It was a blow-out and no wonder! There was not a single tread on the offending tyre, its surface was as smooth as Nibs' nose.

'Oh rass,' cursed the driver, 'I ain't got no spare.' He turned to us and shrugged his shoulders. 'You going have to wait fer anudder taxi to come along. I can't do nothing til tomorrow.'

We hoofed it. Past experience had taught us that it might be hours before another car came along. So we walked the last two miles to Cane Grove while Niblet took his first real swim. There was a long, shallow irrigation canal running parallel to the road and as soon as Niblet saw the flash of water, he strained at his collar. With some difficulty we attached his nylon lead, an extra-long one which allowed us to walk along the side of the canal while Nibs led the way aquatic-style.

It was the first time we had seen him swim. He had been rather aloof, and I think a little afraid, of the bath at the Park Hotel. But the canal was something else, another world altogether. Nibs crept to the edge and dipped his nose in tentatively, testing the dark surface to see if it really was water. Still he hesitated, so Keith nudged him gently with his foot and Nibs slid in with a little plop. In those cool depths he was pure grace, no longer a wriggling, trenchant pet but a creature of the water, twisting and turning with a confidence born of a million years of evolution. I was tempted to set him free there and then but I knew it was not the time or the place to do so. I doubted whether he could fish for himself and even if he could, he would be an easy catch for any hunter. But seeing him frolic happily among the weeds and mud, peering up at us with his big trusting eyes, I made a silent promise to give him his freedom when we felt the conditions right for him. Nothing less would do.

9   The Voice of the Otter

(Liz)

It took us over two weeks to get into any sort of routine with Nibs. He wrought as much change to our world as a first-born baby does to a young couple's. The only difference was that we had not expected the new addition to the family so we weren't sure where we were going to keep him or, for that matter, how we were going to cope with his enormous appetite. At the moment he consumed about a pound of fish every day, which I reckoned we could catch by rod or by Pappy's cast net, but what when his daily intake increased by twice or three times that amount? The two pounds we had bought at Mahaica market (all that would keep without going bad) would give us some breathing space but after that we would have to take to our rods. When we had first arrived in Guyana it had taken us hours to catch a meagre handful of fish, but we had improved considerably over the months, well enough, I hoped, to satisfy Niblet's demands.

Niblet's accommodation caused more of a problem. We had to remember that the rest-house, dilapidated as it was, was not ours to tarnish with the smells and sounds of an otter, so we could not be as easy-going with Nibs as we would have liked. Pappy and Katina, the watchman, were friendly, but they had a job to do and I knew that if we overstepped the mark there would be trouble. With the film and study at stake, it was something we could not afford. The best place for Nibs was where we could keep a close watch on him: our room and the two Victorian showers next to it. This would give him plenty of space to play in while we were out but at the same time would serve to contain his indoor spraint heaps within cleanable boundaries. And later there would be the creek and swamp for him to swim in when one of us was there to chaperone him, so there was little chance of his developing swim-withdrawal symptoms.

It came as no surprise to us that as soon as Niblet was shown his new quarters he lost no time in setting up a loo beneath our sink. We were by now familiar with his tendency to do things at inconvenient times and places, and the choice of toilet was just one of

93

them. The preliminary investigation of room-corners and legged furniture was now old hat to Nibs, a boring routine that had to be executed in order to oblige the rules of otter behaviour. I let him confirm my prediction on the faecal HQ and then installed a tray of sand in the area in an attempt to minimise the smell. It worked beautifully at first, but then Nibs began to expand the spot in ever-increasing circles to the extent that the tray was reduced to a mere showpiece. I threw it away in surrender and resigned myself to shovelling up the dozen or so droppings Nibs deposited every day. We quickly learnt to pick our way through the smelly mine-field in order to reach the sink but we never really got accustomed to cleaning our teeth with spraint-tainted toothpaste. Travelling had taught me that humans can adapt only so far – after that, their only option is tolerance. We were quickly discovering that you had to have a lot of this commodity if you shared your roof with an otter.

Niblet's run extended to the two showers next to our room. They were large squares of tiled cement with a connecting door between them. Nibs sprainted in one and swam in the other. We had to work round his schedule and nip in for a quick shower when he was not around. But the splash of water seldom failed to magic him to our side and the shower that began as a cool relaxation for us ended up as a grand fête for him. If we shut the door to keep him out he would scratch frantically and start up a most unholy barrage of wheep-wheeps, leaving us with the Hobson's choice of mental torture or physical harassment.

Niblet soon learnt his way round Lama base camp. We could not allow him to roam freely on his own in case he established marking posts in unforgivable places such as Pappy's kitchen. He also had a particularly frightening tendency: Niblet, we dis-covered, lacked any concept of air space. To him, it was no less buoyant than water space and therefore no less perilous to plunge into. We got the first inkling of this deficiency one morning as Keith walked him over the land bridge that joined the rest-house's annexe to the main building. Full of beans, Nibs tugged this way and that on his lead, sniffing every sniffable object with great interest. He waddled every now and then to the edge of the planking and peered, with every suggestion of respectful aware-ness, at the ground some eight feet below him. There he went, doing it again for the nth time but suddenly, instead of retreating as before, he launched himself into the air. It took Keith com-pletely by surprise. He could not believe his eyes but the sharp jerk on the lead a fraction of a second later was real enough.

Niblet had not hit the ground but at first sight the alternative looked just as bad. His collar was a noose about his neck and it was this fragile part of his anatomy that had taken the full weight of the fall. In theory, Keith should have been looking at a very dead otter, but Nibs did not deal in theory. He squirmed and wriggled like a maggot on the end of a fishing line, full of life and free of assumptions. Gently, to prevent any further stress on his neck, Keith hauled him up and scooped him back onto terra firma. It had been a traumatic experience for us but not so for the in-domitable victim. He simply shook himself like a dog and trotted off as if nothing at all had happened.

We soon realised that the episode had done nothing to educate Niblet in the dangers of free falling. About a week afterwards, while nosing along the balustrade of the stairhead by our room, Nibs extended his investigations into space and fell seven feet to the landing below. There was a soft thud and I raced, panic-stricken, from the showers where I was just about to call him for his fish. He had fallen on all fours but even so, his little body remained inert on the floor. I went cold, convinced he was dead and blaming myself for not keeping a constant watch on him. I rushed down the stairs and put out my hand to feel for his heartbeat, but his eyes suddenly opened a crack and his little snout moved to one side. 'Thank God!' I said with loud relief. 'Take it easy, little fellow. Easy now, take your time.'

I let him recover slowly without moving him and eventually, after five agonising minutes, he managed to get up and crawl back gingerly upstairs. It took another twenty minutes before he recovered his former exuberance but I was still worried that he might have sustained internal injuries and for the rest of the day kept a close eye on him. I hoped otters had as many lives as cats – Nibs had already used two and at this rate, he would need the full quota.

Nibs never did get the right idea about space, and we learned to keep a sharp watch over him during his periods of free rein indoors. As a result, I learned a great deal about otter behaviour at close quarters and collected some unique data which I hoped would give me an insight into the relationship between the small Guyana otter and the giant otter. Niblet's vocal repertoire was particularly interesting. In keeping with his solitary cousins – those otter species that do not form permanent social groups – his vocabulary was limited to four or five sounds. We were by now familiar with the soft 'ah-ah' gasps he emitted during tickling and swimming and only too well aware of his nerve-shredding wheep-wheeps. This villain of the piece was the species' contact call, a vocalisation used by both young and adult animals to keep in contact with one another. Nibs used it not only to gain our attention but also as a multi-purpose sign of displeasure, hunger or annoyance. Each wheep varied in pitch, loudness and length and Nibs was at his most insistent when all were switched on to 'maximum'. Other *Lutra* species, both in the Old and New World, have a similar contact call though they have never been reported to emit it in such a long repetitive series. The giant otter's contact call is altogether different. It sounds very much like the klaxon horn of a 1920s Ford and can best be described as a two-syllable sneeze-grunt. Often when Keith and I were out in our canoe, we would hear Spotted Dick and Co. make these loud, nasal calls between themselves long before we spotted them.

Nibs was quick to learn the power of the wheep. Every now and then, a commissioner from the Conservancy Board of Directors would visit Lama on a hunting trip and on these red-alert occasions we would endeavour to keep Nibs' profile as low as possible. Experience had taught us that his being seen was not as objectionable to these men of padded conference rooms as was his being heard and smelt. The smells were difficult enough to control, especially within the confines of a building, and capriciously dependent on wind direction. But at least it was possible to scrub floors before the visitors' arrival and remove offending spraint heaps wherever the runways of otter and commissioner coincided. But, short of supplying ear plugs to the commissioners, nothing could be done to combat the sound problem. Out of sight, out of mind is a tidy little aphorism that does not apply to otters. The best we could do was to treat Niblet's wheep-wheeps fatalistically, sticking around Lama whenever commissioners were abroad and taking turns in nipping Niblet's staccatoes in the bud. As a result, Nibs had only to signal his desire for a tickle or a game of chase-the-otter and one of us would fly upstairs to comply. Once he had identified a 'magic' day, he would deploy his weapon with a merciless lack of restraint, bringing us to heel on command. On such days our little otter literally called the tune.

It was to be several weeks before Nibs acquainted us with a completely new call. He had been very good all day, playing happily with some live fish in the bath, but Keith was dying for a shower after coming in from shooting scenics on the river, so I hauled Nibs out and volunteered to distract him for a while. I led him on a root around the rest-house grounds and he took the opportunity to renew his old spraint spots, though by the time he came to number five under the guava tree, he had all but dried up. I then led him to the upstairs verandah of the old annexe to check on some fish skeletons I had left out for the red ants to clean. It was a good spot for the hungry hordes; they came in their thousands from a nest nearby and were very efficient in removing every last piece of flesh from my morbid display of bodies, leaving skull and skeleton beautifully intact. Nibs had never been here before. I had designated it an otter-free zone as it was the only place I could leave spraint samples and other tools of my trade lying around in relative safety. Nibs trotted in without a moment's hesitation when I opened the termite-ridden door, and set about sniffing every corner and cobweb.

As I poked at my whitening fish-bones, I suddenly became aware of a low chuttering sound. I looked up and saw Nibs trotting hurriedly back and forth across the room, head held low. The noise was coming from him. He made it with his mouth shut and, when I listened carefully, it seemed to emanate from his nose rather than his throat. He looked patently anxious and I bent down to comfort him but he avoided my hand and continued to pace the room and make his muffled rattle. He sprainted repeatedly, or at least tried to, in one spot under an old table, but this determined show of ownership did nothing to lift his apprehension. It was the first time I had known him to refuse the offer of a belly-tickle and a finger-bite. In the end, I had to take him back over to his familiar territory with its cherished swimming pool and familiar spraint heaps.

Nibs' new call was clearly the well-documented 'chuckle' that most *Lutra* species give when greeting one another, the so-called 'close contact', or 'affectional greeting' call. It was obvious, though, that Nibs had not used it in that context upstairs on the verandah nor in the numerous other situations that we witnessed. He always chuckled in a situation of mild stress: when he badly wanted a drink of water, when he wanted to cool himself off in the shower but could not, in certain unfamiliar places and once when he caught me removing one of his spraint heaps. In all these situations, he was consistently tense and would refuse to be distracted by petting or play. Sometimes we could not or would not comply with his wishes and he would then bring out the big guns, flailing our senses with wheep after wheep. Drawing conclusions about animal behaviour in the wild from captive observations is always a risky business, and I knew I could not be absolutely certain that Niblet's wild-living peers used their chuckles in the same way. Certainly, Nibs differed from his other *Lutra* relatives in that he never used it as a 'hello' signal, even when he greeted and touched noses with the dogs at Lama.

I was in for another surprise with Niblet's fourth sound. It was a cross between a moan and a sigh, more appropriate to a pissed-off human than an otter cub. The noise lasted about three or four seconds and was never repeated. Nibs kept his mouth closed while making it and this gave us the uncanny impression of a ventriloquist using our otter as a puppet. It was a sound Nibs used very conservatively and we heard it only about every other day. Moaning has also been described in the North American otter, *Lutra canadensis*, in situations of mild alarm. But Nibs never used the chuckle for this purpose, the moans actually being given in very relaxed situations: during grooming, shortly before taking a nap or during a lull in his games with us. The only other times we heard him moan was while he was doing something which zoologists have never believed otters do but which local hunters and fisherman have always insisted they have seen them do: storing fish.

The first we knew of Nibs' penchant for storing fish was one afternoon while taking some stills of him in the bath. Each couple of shots qualified him for a reward of one live fish so he was in a very co-operative mood. Already he had consumed about ten good-sized fish in as many minutes and the fuller he got, the more he played with them. He would chase the poor creatures to exhaustion and when they finally collapsed and floated to the surface, he endeavoured to nudge them back to life with his snout. Supplied with his eleventh patwa he proceeded to repeat the act, ending by throwing it up in the air so that it slid down his left thigh and rubbing it vigorously between his forepaws while lying on his back in the shallows. Suddenly, without warning, he clambered out of the bath with the four-inch fish in his jaws and walked past us to the adjoining toilet. He disappeared round the half-open door and through the crack we saw him drop the lifeless creature in the corner and release a long bass moan. He returned to the bath as if nothing had happened and set about dealing with the next fish.

After that, Nibs got into the habit of storing fish in our room. His favourite pantry was the corner behind the anti-fire sand bucket, though we sometimes found the odd half-chewed specimen under our beds. We would give him a late-evening snack whenever our fish supplies permitted. If after satisfying his appetite there were a couple of fish left, he would dutifully trot over to the sand bucket and drop them in the corner, with or without an accompanying groan. They were always gone the next morning. Rarely did he store food if he was still hungry but theroomas – small nocturnal catfish – were the one exception to the rule. Theroomas were the otter equivalent of 'greens' – eaten when necessary. They were all too often the only fish we could catch

98

and we always carefully removed their bony heads and pectoral spines before feeding them to our baby carnivore.

There was one day when we managed to catch only six of his favourite flying patwa and four theroomas for his supper. He normally disposed of a dozen or more one-ounce fish and that evening he polished off the six patwa in quick succession. He then turned to the theroomas, recognised them for what they were – the lowest form of otter food – and nibbled one of them unenthusiastically. There was clearly still an empty space in his stomach which he wanted to fill, but he could not bring himself to do so with such repulsive morsels. All four specimens were transferred to the fish-safe behind the sand bucket and we would have forgotten about them had not a ghastly odour pervaded the room days later. With theroomas, Nibs' fish-piling behaviour amounted to disposal rather than storage.

Only once did I invoke the wrath of Nibs' fear-attack cry and I would not wish to do so again. Keith and I had just got into bed and everything was in darkness when I realised I had left the mosquito repellent in the showers where Nibs was sleeping. Without it, nights at Lama were long tunnels of itchy insomnia so I had to go and get it. My torch-light fell on Nibs as I opened the door. He was curled up cosily among the folds of his frayed blanket. His head moved slightly as I entered. A fraction of a second later, he was awake and had turned to face me, his eyes reflecting two green lights, his little body stiffening in fear, head lowered, forefeet spread out, hackles up like a dog. 'Gaaaasshhh!' The feline hiss shot from his open jaws and warned me to stay exactly where I was. At that moment I feared him as much as he feared me. He looked up, ominously ready to pounce, and I realised that there was no sense in testing the defence capabilities of a frightened otter, pet or no pet. Reassuring him gently that it was only me and that everything was all right, I gave him the time he needed to clear his sleep-befuddled brain. He eventually relaxed and allowed me to pet him back to sleep, but I resolved there and then never to pay him a surprise visit again during the night.

What struck me about Niblet's calls was their distinctness from one another. A giant otter's vocabulary is full of variations that make the identification of distinct calls a somewhat arbitrary business. The difference in the two vocabularies is probably a reflection of social organisation. Solitary species like the small Guyana otter do not have as much need for a large and complex range of sounds as do social species whose members have to communicate with one another more frequently and more subtly to maintain social cohesion. Spotted Dick's family provided me with the full gamut of graded sounds and from these I managed to categorise eight distinct calls, many of which were shared by the captive female in Georgetown Zoo. 'Hahs!' were a sign of mild alarm, snorts being invoked only when there was more to be worried about. The most common call was the sneeze-grunt which served as a kind of vocal bond, the contact call that kept family members close together. Without it, group cohesion would have been a shambles. Caterwauling was the weirdest call of the lot. When we first heard it, I thought we were approaching some over-burdened nursery of new-born babies hidden away in the jungle. It expressed all the pent-up frustrations of an otter that couldn't get what it wanted. Needless to say, it was more characteristic of young cubs, who lacked the hunting prowess but not the appetite of their parents. We often saw them having to suffer the agonising sight of fish after delicious fish disappearing down mum or dad's throat without the smallest hope of a morsel coming their way.

The adults, and later the cubs themselves, reinforced their hands-off attitude with low hum-growls of warning: kin they may be but where food was concerned self-interest was the name of the game. Short squeals and dog-like barks also figured in the giant otter repertoire but their functions were never very clear. On the other hand, screams conveyed an obvious message. These terrible suggestions of torture were only ever given by a family group and accompanied sham attacks, directed exclusively towards us. As a war cry, they were superbly effective and put the fear of God into us the first time they rushed our boat. It was only after the third such encounter that we realised it was all bluff. Then and only then could we afford to sit back and watch the charges and listen to the screams as we would a House of Hammer horror movie – with a feeling of nice, secure excitement.

For all its fearsome moments, recording and analysing wild and captive animal calls is one of the cleaner aspects of an ecologist's trade. Collecting and sorting through droppings is not. My days consisted not only of scooping up giant otter spraints – quite often fresh – but also of cleaning up Niblet's which were always fresh. It was just as well I had more than a hygienic interest in Nibs' spraint. It was important for me to find out whether or not sorting through otter droppings to identify the undigested bits and pieces was a good way of finding out what the animals actually ate and in what proportions. To do this, I kept a daily record of the number, size and type of fish we fed Nibs, and collected all the resulting spraint the next morning before feeding him. I planned later on to carry out a similar experiment with Squeaky, the captive giant otter in Georgetown Zoo. Every day I dutifully washed my samples in small bags of fine muslin and hung them up to dry. I was always amazed at how I never thought twice about doing this, yet to wash our own clothes created agonies of procrastination. But I knew that, despite its apparent absurdity, rinsing otter excrement was far more worthwhile than scrubbing clothes. The first helped unravel a mystery whereas the second did nothing more than mark the beginning of yet another cycle of dirt.

It was surprising the traffic of scales, vertebrae, fin rays, spines, bony plates and fat (when we fed Nibs mince) that passed through Niblet's tiny gut. These remnants enabled me to analyse Niblet's meals with remarkable accuracy. The experiment had proved the worth of droppings as an index of otter diet and it meant that I could now use the method with confidence to discover what Niblet's confrères ate in the wild. It was time to seek out these elusive dwarfs which Ram Raj the hunter boasted were so plentiful around Shank's Canal.

We made an early start and set off just after dawn one morning. We had decided to take Niblet with us. This would eliminate many of the time-consuming morning chores that were part and parcel of keeping an otter: feeding, changing bath water, replacing grooming mat with dry one, giving half an hour's play, dropping live fish into bath for otter to play with – the list was endless. Taking Niblet would also give me the chance to see his reactions to the wild spraint I hoped we would find.

Most days, we tried to avoid bringing Nibs with us in the boat because his antics demanded so much attention. All went well as long as he swam beside us on his lead but unfortunately Nibs did not stick to straightforward movements. After a few minutes of model behaviour he would start to chase fish under the lily mats and dive down to explore the submerged takoubas, or rotten logs. It took our combined efforts to watch that his lead did not snag on anything, and if it did to undo the tangle before he ran out of air. Preventing him from swimming demanded even more energy and produced instant vocal protest. On top of all this we had to maintain a steady course towards our destination along a route fraught with sudden bends and narrow channels. Two such trips and one aborted fiasco later, we were convinced that pet and project were incompatible. There was no way we could let Niblet's pleasure override the interests of the film and the study so Nibs was left behind at base camp on the trips that required

101

concentration and quiet. So that we did not feel too guilty on these occasions, we would be sure to leave him a dozen or so live fish to pursue and play with in his bath pool. And he was also guaranteed our undivided attention and a swim in the river when we returned.

But the expedition to look for signs of wild *Lutra enudris* was different. We wanted Nibs to accompany us to help us find the small, untrampled spraint heaps these little otters made and to register his reactions.

We decided to investigate Shank's Island first. Ram had described the area as a remote waterway of small floating islands and boggy swamps to the north-west of Russel Lake. It had looked quite accessible on the map but thick mats of weed and water lilies hindered our progress and it was midday by the time we arrived at the first floating island. Nibs could not stand the heat of the boat, despite the three or four inches of 'leak' water that rolled round our feet, and he spent most of the time in the lake, alternately swimming and being pulled. This hitch-a-ride game was a very recent invention of his. He had arrived at it more by accident than design during the fifteen-mile journey and he now revelled in it, allowing himself to be towed belly up or down. The only effort on his part was to raise his snout every now and then for a lungful of air. The game pleased us just as much because it meant we did not have to be constantly on the look-out for potential lead-snaggers.

We tied up on a small spit of dry land that was barely four inches from being inundated. A pair of flycatchers fluttered into view and performed a beautiful pas de deux over the water. They rose and fell in delicate, fluttering movements and then disappeared across the sedge without waiting for an encore. Nibs tugged on his lead as we clambered out to start the survey, wild with joy at the sight and smell of the grass clumps, low bushes and earth mounds that lay around him, all free for the sniffing and sprainting. We watched as he squatted seven times within the space of about three or four minutes, though the fact that his last attempt lacked any sort of end-product deterred him not one bit. His ottery make-up constrained him to mark out a territory and to leave messages, and this he did with every sign of enjoyment, quite unaware that some bulletins never got through. The ritual was all.

There was only one way we could go and that was towards the knuckle of land whose finger we had moored on. It was the perfect jetty, running uninterrupted for about half a mile across the still water before opening up into a broad expanse of seasonal savannah swamp. At the height of the dry season this type of swamp often loses three to four feet of its water to expose a floor of sun-baked mud between the cutting grass and Ite palms. We could always tell a seasonal savannah swamp from this character-istic plant cover. Patches of broadleaf trees and shrubs, indicating unfloodable land, pin-pointed a more likely surface for otter scent posts. Nibs led the way spasmodically. He would pull at the lead and then get so engrossed in some invisible scent that we would have to overtake him and tug at his collar to coax him along. The thin strip of land was little more than an arm's length across at its widest and in places the water pinched it into wasp-like waists. Progress was slow at times. Thick clumps of stunted trees and woody shrubs tore at our skin and we had to protect our eyes with cupped hands. Once I nearly grabbed hold of a piece of living glass – an orange-green 'hairy worm' – which put the fear of God into me and made me curse at my self-imposed predicament. These fine-haired caterpillars came in all sizes, colours and hair lengths, and contact with the poison-drenched filaments meant sure hospitalisation and days of pain. So professional was their camouflage that I was surprised we had not fallen victim to one already.

Nibs was immune to the scratch-and-sting tangle. His little body easily fitted underneath it all and he trotted along in com-fort, taking advantage of our numerous pauses to sniff the undergrowth and leave a token deposit whenever possible. About a third of the way along the promontory his sniffs suddenly became more intense and he pulled Keith towards a bare patch of ground in the middle of a short length of open grass. Two weathered spraints lay side by side on a small mound of com-pacted earth, the result of generations of such deposits. 'Otters,' said Keith, and I nodded in excitement. The scales and bones were obvious on the surface and there was a distinct smell of otter when I stooped down for a sniff. Had they been fresh, we prob-ably would have smelt them a hundred yards away but they were about a week old, though still dark in colour and still retaining their cylindrical shape. Another week or so and they would start disintegrating into grey heaps of fish fragments, the fire ants having first helped the sun and rain to dispose of the soft con-tents.

Nibs could not wait to read the two messages and deposit his own alongside. I wished I could have read scents as easily for I would have discovered the ages of the depositors, their sex, their breeding condition – whether in heat or not – and their status in the local population. As it was, I had to content myself with measuring the diameter of the spraints to confirm that they were the work of *Lutra enudris*. Giant otter droppings are thicker than those produced by the smaller species, though the spraints must be fresh or at least intact in order for the half-centimetre difference to be measured. Not wishing to remove the scent messages completely for fear I would disrupt otter communication in the area, I took half of each spraint and dropped it in one of my sample bags for later analysis. It was important to find out if Niblet's wild relatives ate the same species and sizes of fish as the giant otter. My hunch was that there would be a difference but with some overlap.

Half a dozen paces further on, we came upon another mound which looked as if it had not been marked for several weeks. The scale and bone fragments had been whitened by the sun and I could not detect the faintest smell even when I brought handfuls of the loose remains to my nose. It was quite clear, though, that Nibs did not share my human insensitivity. He nosed the area appreciatively and responded to the whisper of scent by expelling a small blob of gut mucus, all he could spare now that the roughage had run out. Keith grinned. 'Better go easy on the ink, old boy. If not, you'll have nothing left to write with!'

We were well acquainted with Nibs' acute sense of smell. One day at Lama I ran out of dry grooming mats and had to turn to an old towel I used for cleaning up his spraint heaps. It had been washed thoroughly in detergent, soaked in disinfectant, rinsed for a good twenty minutes and hung out to dry in the sun. Joking with Keith, I acted the happy housewife of TV commercial fame and gave it the 'close nose test'. Not a whiff of otter, only a satisfying washday perfume, and yet, to my horror, Nibs promptly sprainted all over it! He had no problem in detecting his scent beneath the layers of artificial odours. He also responded without hesitation to well-weathered giant otter spraints deposited weeks before on the river banks around Lama.

We reached the end of the promontory without finding any further evidence of the Guyana otter. The irregular circle of the savannah swamp spread out before us, a skin of pale grass veined with the dark green of tree-covered dry land. The rainy season was not due to begin for another month yet, but a spate of freak showers the week before had caused a premature filling of the swamp. At close quarters, the water showed black in between the dense clumps of grass. We searched every square inch of dry land and found nothing. There was only one area left to search and that was a small hummock of land completely surrounded by swamp. Between it and us lay eight feet of black water. If we had not had Nibs to control, we would have tried jumping it, but we could not risk losing him so we were condemned to swim or wade depending on how deep the water was. Keith cautiously submerged a rubber boot and slowly shifted his weight from the dry one. Another step into the swamp and he was up to his armpits in water. 'Oh Christ! Grab me, Liz, I'm sinking!'

'Get away, Niblet!' I yelled. He was cavorting in the water around Keith, convinced that it was all a new game we were staging just for his benefit. I dragged him in with one hand and reached out to Keith with the other. The water had risen to his neck and his efforts to help himself only drove him further into the soft mud. I sat back and dug into the ground with my heels, yanking at Keith under the armpits. After five agonising minutes, he managed to grasp a tree root and with his help I managed to pull him free, his feet coming out of the wellies with triumphant gurgles. Water gushed in torrents from the large thigh-high pocket on his ex-army trousers. Totally indifferent to the gravity of the situation, Nibs immediately pounced on the exposed toes, causing their owner to protest loudly. I hauled at our indefatigable pet before he got what he was asking for from my post-crisis husband and suggested we swim it this time. 'It'd save us a lot of hassle.'

It did, but it was not much pleasanter than Keith's experience for the cutting grass sliced our skin and obstructed progress and we had to keep Nibs on a short lead for fear of entanglement. But it was worth it. We found another mound with two very fresh spraints poised on top. Nibs strained to add his signature to the little scrolls but failure looked imminent. In a last-ditch effort, he re-squatted over them and, with a look of deep concentration, farted. He then walked up to me, the scapegoat, and nibbled the arch of my foot in frustration. He had done this at Lama as a 'leave-me-alone' sign and sometimes in agitation when he could not go somewhere he wanted. Distraction was the only immediate cure so I led him over to the water where he quickly forgot his tensions in a series of dives and water sports.

I was extremely pleased with the day's findings. A count of three spraint heaps on my data sheets was ample evidence for the presence of the small Guyana otter in the Demerara swamps. Later we were also to find ample evidence of *Lutra enudris* on the agricultural dykes around Shank's Canal. Ram Raj had not exaggerated when he said Niblet's wild cousins were well established among the rice paddies and sugar-cane fields. Interestingly enough, though, there appeared to be no giant otters whatever in the area and we found the reverse situation in the river wolf's core areas and overlap zones around Lama. This segregation implied that the two otter species were trying to keep out of each other's way in order to avoid serious competition for food and dry land. But there was at least one area we discovered they *did* share. This was where Shank's Canal flowed into the northern waterway of Russel Lake. It was used by a pair of *L. enudris* and a giant otter family of four. Neither species used the adjoining north dyke for marking on but there may well have been competition for food. If there was any argument over this resource, then I expected that *L. enudris* would do most of the modifying. It was obviously the more flexible of the two species as it lived in areas close to human habitation and regions which did not have a lot of shade.

I found this remarkably similar to the relationship between mink and otters in Europe: the otters there prefer the well-stocked lowland streams and effectively keep out peripheral mink populations. There is competition in the overlap zone, especially in the winter, but a serious clash is avoided by the mink shifting its diet to include more terrestrial prey. In the same way, where the Davids and Goliaths of Russel Lake overlapped, competition was most likely minimised by the smaller otter eating different prey species and having a different time-table of activity. My sightings of *L. enudris* were sparse so I could not really say when it tended to be most active, but in those areas they shared with giant otters in Surinam, Nicole Duplaix found them to be out and about most often around dusk. However, on the diet strategy, I had rather more to offer. Our experience with Nibs suggested that this way of avoiding competition was feasible. We had known for a long time that Nibs liked hunting for crevice-dwelling invertebrates, spending hours scrabbling among the bank holes in a manner we had never seen Squeaky or any wild giant otter do. But even more exceptional was his ability to hunt successfully on land. He twice caught lizards on the bank close to the rest-house and I also found reptile remains in two of the wild spraints I had sampled around Shank's Canal. Unlike the biblical story, this David seemed more willing to compromise with Goliath than to fight him.

106

10   Patterns

(Liz)

One big surprise came our way the following week. For the eight months of our acquaintance with Spotted Dick's family, we had assumed that they patrolled Lama Pond as well as Lama Creek. Lama Pond was the name we gave to the large area of open water in Russel Lake close to the rest-house. We would often see the family foursome fishing in it or hear them caterwauling among the reed beds that rimmed its perimeter. We had never actually seen their chest-markings because they always disappeared as soon as they saw us cast off from the boat-house.

One morning Pappy interrupted our breakfast to tell us he heard water dogs playing among the sedge. 'I see dem come from Maduni side. Dey cross de weeds by de big palm and den disappear, but dey not far off!' He and Katina were by now quite used to our mad desire to shoot the beasts with our cameras and they would give us a shout whenever they saw or heard anything ottery in the vicinity. Keith and I dropped our buttered rotis and grabbed the camera equipment.

There was no point in using the boat again as we would probably end up filming empty ripples for the third time. We decided instead to make for the nearest cross-over point on the east dyke, a well-used otter runway that linked Russel Lake with the seasonal swamps. It was situated under an old mango tree about half a mile from base camp. There were no other cross-over points for another mile or so in either direction along the dyke so I reckoned there was as good a chance as any that Spotted Dick and his family would use it that morning to reach Lama Creek. We would try to film them as they approached the cross-over from the water, but the shots of them actually clambering over the land bridge would have to wait until we could set up a hide. A sudden klaxon from among the lily mats gave us added impetus – the otters were very close. Keith screwed the Arriflex onto the tripod and set the zoom to wide-angle, while I took the light reading and adjusted the aperture for him. The stills camera was already focussed up and ready to fire.

Five minutes went by, then fifteen and then half an hour. Still no sign of the family; there was not even the odd splash or caterwaul that usually marked their progress – just complete silence. Where had they gone? Were they still keeping their course towards us? A pair of hoatzins suddenly squawked in alarm. We were well acquainted with this bird's weakness for overacting and kicking up a fuss at the slightest provocation, but this was one time we blessed them for it, for they had pinpointed the group's position behind a small clump of mock paw-paws about a hundred yards into the lake. I was getting a little impatient with the otters' long silence and it suddenly struck me that they must have heard us on the dyke and were a little apprehensive about using the land bridge. I motioned to Keith that I was going to return to the rest-house and did so with a fair show of noise so that they would think the human danger had gone. It was the old bird-photographer's trick in which two people enter a hide and one comes back out. Not being able to count, the birds think that the threat has gone, and they relax and go about their usual business. I would not have to forfeit the chance of collecting behavioural data because I could always analyse any film that Keith took later on. I hummed my way back to Lama, crawling my way through the undergrowth and making as much noise as possible, and waited impatiently for the outcome.

Keith burst through the bamboo grove an hour and a half later, grinning from ear to ear. 'Hey, Liz, it worked, our ploy worked! I've got 'em on film! But it's not Spotted Dick and Co., it's an entirely different group!' He patted both cameras with satisfaction and filled me in on what had happened. Within minutes of my leaving, a head had appeared from among the water lilies about twenty feet from the dyke. It started to swim across the open waterway but then it suddenly saw Keith and stopped. 'It looked so surprised to see me,' said Keith. 'Couldn't understand how it was they had heard me go and yet I was still there.' Another head had surfaced alongside, followed by two others. All four of them then started to periscope and it was only after shooting off half the film that Keith realised their chest-markings were different from those of Spotted Dick and Co.'s. We called the parents Mister and Missus and the cubs Dot and Scratch, but referred to them collectively as the Streaky group on account of Mister's complex chest pattern.

Curiously enough, Keith had noticed a marked difference in swimming behaviour between Spotted Dick's family and the Streaky group. Whereas Spotted Dick and Co. always approached our canoe in a noisy two-plus-two formation of parents

in front and cubs behind, the Streaky group had arranged themselves in a straight line and periscoped in total silence, without so
much as a short 'hah!' of surprise. It was all very exciting but it was
also a warning for me to be more careful in identifying giant otter
groups.

Once I realised we were dealing with two groups, the sightings
of Spotted Dick's family in Lama Creek suddenly fell into a
definite pattern. The group arrived there every two weeks or so
and remained for two to three days each time. Only once was
there a major hiccup in the cycle and that was when, for some
reason, they extended their visit to Lama for a record eleven days.
It was impossible to discover where they went during their fortnightly absences from Lama. They just seemed to vanish. To use
our outboard motor to follow them was out of the question
because it scared them away immediately. So we had to resort to
paddle and muscle-power. It was exhausting work to try to match
even their slowest speed and normally we lost them within the
first three miles. Our record 'contact' lasted from the top of Lama
to within 200 yards of its junction with the Mahaica, a distance of
five miles. Spotted Dick had just finished off a patwa behind some
moca-moca and he emerged with Black Throat and the cubs in
tow to find us still hanging around. He gave a long caterwaul of
irritation and headed downstream with his retinue, keeping close
to the left bank. They disappeared abruptly into some overhanging vegetation and that was the last we saw of them on that
occasion. It was no use, I thought, when it came to eluding nosey
scientists these creatures were without peer.

We never once saw Spotted Dick's group along the Mahaica
River, in Maduni Creek or in Russel Lake, and search as we might
along side-streams for marking sites, couches and other indirect
evidence, we found not the slightest trace in these areas. The only
other hide-out was the seasonal swamp that stretched out on
either side of the creek behind the ribbon of riverine forest. We
found from experience that these vast expanses of cutting grass
and Ite palm could hardly be called 'seasonal' swamps. They were
certainly quite unlike Russel Lake but although the water level
fluctuated between two and five feet, they were seldom free of
water and probably provided a very convenient larder for the
local giant otters. Perhaps Spotted Dick's group hunted there
when they were not in Lama Creek. The swamps were quite
inaccessible to humans and there were enough secluded islands to
satisfy the most ardent sprainter. The only disturbance came in
the shape of two subsistence farmers who had clearings on two of
the more accessible islands but apart from this very localised

interference, the otters had complete privacy. To get to their smallholdings the farmers had each felled a dozen or so Ite palms and had placed them end to end over the swamp. Being buoyant, they made marvellous, if rather unsteady, bridges. Unfortunately, it was not a technique I could use to reach all the islands I wanted to examine, as the disturbance would have been far too great. I thought of using stilts to survey the area but I quickly discarded the idea after testing the ground with a long pole and finding that it sank all too easily into the mud. I had to content myself with studying the comings and goings of the family groups within the confines of Russel Lake and the creeks.

It became clear that the Streaky family was also following a definite travel pattern. They visited Lama Pond every three weeks or so, staying over for about three days to fish and mark. Because their timetables were only approximate, it sometimes happened that both groups visited their core areas together. The two properties were separated only by the rest-house and the twelve-foot wide east dyke so this was probably the closest the families ever got to each other. Double visits were always times of hectic activity for me. I would get up at dawn to observe Spotted Dick's early-rising family in Lama Creek and then, at around eight-thirty, cross the dyke into Russel Lake to meet the Streaky group. I often ended up eating breakfast at eleven o'clock.

We found the same sort of movement pattern in two other family groups we discovered shortly afterwards in the vicinity of Russel Lake. This brought our total 'score' to four groups: the Maduni group of four, whose core encompassed Maduni Creek, and the Anadale threesome, which owned the narrow waterways of Anira Creek and Anadale Gutter. Tumbling to their cyclical pattern of travel was the biggest breakthrough of the study because it allowed us to predict the arrival time of each group. About three weeks after Keith had first met the Streaky family, we paddled across to Lama Pond and lay in wait for them among some scraggy bushes. We struck lucky on the second day out. The foursome appeared suddenly from among the lilies and moca-moca opposite our hide-out and spotted us almost immediately. Looking very surprised to see us, they periscoped earnestly for a better view but did not snort or vocalise in any way. After a few minutes of cobra-like swaying, they sank back down and disappeared. The following day they were expecting us and barely peeped above the surface of the water before making another silent exit.

Spotted Dick and his family were exhibitionists by comparison, and their visits to Lama gave me a good idea of what a day in the

110

life of a giant otter family was all about. They divided the twelve hours or so of daylight into fishing, travelling and land activities. Hunting and travelling consumed a good two-thirds of the day, while grooming, resting and defence duties took up the rest. Nothing was more exciting (or uncomfortable!) than watching a giant otter performance on the sly. A camouflaged hide on the bankside in Dark Lane gave us a fairly clear view of one site's entrances and latrines. It was Spotted Dick and Co.'s favourite, the one they used most often on their fortnightly visits to Lama Creek. We became past-masters at squashing mosquitoes in silence, an art that proved more difficult than it sounds because it required both vocal restraint and a peculiar kind of slow-motion slap. It was not that we had come unprepared. We knew we would get a nasty reception and had erected a mosquito net over the hide; but with a Ninja-like fanaticism, a good twenty per cent of the biting hordes still managed to find a way through. The whine-and-bite torture had, however, paid off during the first few weeks at Lama and one memorable afternoon we were rewarded by our first close-up view of giant otter behaviour on land.

Spotted Dick was the first to clamber out of the water. He made straight for the current latrine high up on the bank, sniffed it thoroughly and then turned round to spraint and urinate, his tail held high. It was gratifying to see the whole animal – all four legs plus tail and torso – and to reassure myself that I had not just been chasing floating heads during those first anxious weeks of the study. Spotted Dick waved his tail vigorously up and down after he had finished and gave the spraint a good tamping down with his hind legs. It was a stiff, rocking movement that soon trans-mitted itself to his forelegs. He arched back and forth over the dung heap and at the same time bobbed his head up and down energetically. Black Throat then decided it was her turn to ablute and as soon as she climbed out of the water her mate disappeared among the foliage. Although we could not see him clearly, I could just make out his dark shape stretched up in tripod stance, pulling down small branches and low saplings. He rubbed the leaves with vigorous circular movements between his forepaws and down his chest and as he fell back down on all fours, he trod the whole lot into the mud. Black Throat then joined in on the opposite side of the site. We could not see her but the sound of swishing leaves and the flip-flip of her webs on the soft mud came over clearly. She was the first to slip back into the water where her cubs had been waiting all this time. Spotted Dick joined them shortly afterwards and the foursome slowly headed back down Dark Lane to the open creek.

111

This first sighting of Spotted Dick and Black Throat on land was the culmination of weeks of hard slog and hopeful persistence. It marked the first crucial exam in our long course on how to be an otter voyeur. I learnt a great deal from the encounter, brief as it was. For one thing, it was Spotted Dick and not Black Throat who had been the first to land on the site and mark. I later discovered that Nicole Duplaix had found the opposite. The female of the giant otter pair she watched in Surinam was always the first to initiate land activities such as marking, gardening and grooming, whereas the male concentrated on defending the family. Nicole concluded from this that female giant otters are dominant over their male partners, a role inversion that most feminists would be delighted to quote. Unfortunately, Spotted Dick and Black Throat complicated the neat picture of female dominance.

But what was at least as interesting to me was the cubs' failure to venture onto the site. They just mooched around in the shallows, leaving all the sprainting and gardening to their parents. Why, I wondered, hadn't they joined in? Was marking perhaps a fairly high-status activity confined to adults? Ringo and her brother were only about four months old at the time so perhaps it was a case of children being seen but not smelt! And there was indirect evidence that the same thing had occurred with the Maduni family whose two cubs, Tom and Jerry, were about the same age as Spotted Dick's. Shortly before dusk one evening, I found two sets of adult spoor and tail-drag marks at the entrance of a very freshly re-marked site. I knew the family was in residence at the time because we had sighted them off and on during the day in the headwaters of the creek not far from Iburu's hut. No other pairs or groups were in the area so the site could only have been visited by Tom and Jerry's parents. The young ones must have waited in the shallows until they had finished. I later learnt that Nicole Duplaix had seen the same thing with a family of three in Surinam. The parents 'came ashore separately to defecate, the female first followed by the male 4 minutes later. They defecated, urinated and spread their faeces 14 minutes later in the same place and again the male followed the female, this time immediately afterwards. The juvenile did not follow its parents ashore on either occasion but waited for them in the shallows at the base of the site.' It seemed that when it came to leaving messages, giant otter cubs were excluded from communal participation. Their young spraint, probably as uninformative as childish scribbles are to adult humans, might seriously scramble communication between adults. Sub-adult spraint was more than likely just wasted in the water. But the arrangement was not a permanent one, as I was to discover several months later.

In mid-April 1980, around Ringo and Spotted Dick Jr's first birthday, I discovered that the foursome had begun to mark *en famille*. To do this, I once again turned to spraint for help and to some earlier work I had done with Squeaky at Georgetown Zoo. I had weighed lots of washed and air-dried droppings of Squeaky's and had found the average weight of a single spraint. I then cleared the current latrines on each of the Dark Lane sites, leaving only a small portion to guard against totally disrupting the family's communication system. I monitored the sites regularly and whenever Spotted Dick and his family visited one of them, I scraped up the entire deposit, again taking care to leave half a handful or so as a stimulus label. The next step was to wash, air-dry and weigh the smelly bundle and to divide the figure by the weight of a single spraint. This gave me the number of droppings that had gone into one marking session. Looking back at my Squeaky notes, I found that the average number of deposits she had made during each spranting session was roughly one. It followed that the number of droppings in each Dark Lane latrine was therefore the same as the number of otters that had sprainted. It was a laborious way of gathering information but with a study subject as shy as mine, it was the only way. I had always to keep my imagination on the boil in order to devise strategies for extracting their secrets. No idea was too mad to be considered, no plan too outlandish. Ecology, I was beginning to realise, was much more than stoic use of eyes and ears; it often bordered on an unlikely brand of forensic science.

Nevertheless, it got me what I wanted. All through January and February 1980 I realised that only two otters – Spotted Dick and Black Throat – were doing the sprainting on the latrines in Dark Lane. Early in March, however, the quantity of spraint suddenly increased and I calculated that more than two otters were now contributing to the latrines. Ringo and Spotted Dick Jr were growing up. They had graduated from infant potty to adult loo and as qualified site-markers were now allowed to blend their unique brand of scent with their parents' to produce a composite family smell.

It was not long before this remarkable shift in cub status bore fruit. It had an almost immediate impact on single young adults that occasionally passed through the Spotted Dick family territory. The travel movements of these transients suddenly changed from a random pattern to a periodic one and the new cyclical itinerary followed an intriguing pattern. The connection did not hit me straightaway and it was a good six weeks before I realised what was going on. I looked at my data sheets and saw that from early March, sightings of the single transient in Lama Creek were made either during Spotted Dick and Co.'s fortnightly visits or a day or two afterwards. It was a clear case of follow-the-group, a pattern which gained further significance because of the single, untrampled spraints I began to find on Spotted Dick's marking sites. After every group-plus-transient circuit I would find a single intact dropping, set apart from the refurbished family loo. It was usually slightly fresher than the communal heap and this meant that it had been deposited a day or so later. It suddenly became obvious that the single spraints were the work of the transient: he or she had read the latrine smells, realised there were two rapidly maturing year-old cubs around and decided to keep close tabs on them. This finding came as something of a surprise as Nicole Duplaix had never once found single spraints on communal marking sites in Surinam.

As it turned out, Shadow, as I called the transient, followed the group for just over a year, by which time Ringo and Spotted Dick Jr were ready to cut their apron strings. Shadow's marking activity suddenly intensified in the last two visits before the family's disintegration, when he (or she) not only placed his routine blob of spraint next to the family pile but also made three sites of his own down Dark Lane. These were well spaced and well trampled but, interestingly enough, the single spraints were still never trodden into the ground. I remembered that Dancer, the single female we had watched on the east dyke, had done the same thing and I wondered if the stimulus of a partner was a necessary ingredient for spraint-mashing. If the purpose of dung-mixing was to blend the partners' smell in order to create a composite scent (as the sign of a strong pair-bond), then it made no sense for a transient to knead its faeces.

For the entire year that Shadow followed Spotted Dick's family, he rarely appeared on Lama other than when the foursome were in residence. Once I was in a position to predict his arrival, I usually kept an eye out for him. He would appear first in the Streaky group's territory – either in the south canal or in Lama Pond – and would cross over to Lama Creek via the mango tree cross-over. From the lack of single spraints on any of the Streaky group's marking sites, it was clear that the Lama transient was not particularly interested in the Streaky cubs even though their visits often coincided. There was no doubt, though, about the Anadale family having their own silent satellite. The small, diagnostic messages generally appeared after the group visited their site in

Anadale Gutter but the transient seemed shyer here and was never as faithful as Shadow was to Spotted Dick and Co.

Shyness was something our capuchin friends had never heard of. Big Boy and his troop were back at Lama Creek after an absence of several weeks. Their human-like expressions and cute antics took me out of my preoccupations with transients and single spraints and provided me with the entertainment I needed. One day we met them jumping a leafy bridge high up across the creek. Craning our necks, we watched as each one in turn sprang over the chasm and fell with a loud rustle onto the other side. There were two mothers with young, one a tiny, black-headed mite clinging to her belly, the other a juvenile riding piggy-back. Not for an instant did either hesitate to leap the gap. It was a routine part of monkey travel. We followed them down the creek, where our persistent gazing eventually brought out the warrior in Big Boy. He started to do his branch-shaking act, trying at the same time to stare us into retreat. Every now and then he took a break and stuffed himself with figs. During one of these intervals, as he was reaching for a particularly succulent bunch of fruit, a blurred object suddenly hurtled out of the sky. The capuchin troop erupted into panic-stricken chitters. But their warning came too late for Big Boy. With a shrill scream, the eagle swooped through the canopy and hit Big Boy with more than a hundred pounds of power. Its talons struck home and ripped the leader of the troop off his perch.

Knocked unconscious by the violence of the strike, Big Boy hung lifeless in the predator's talons, a limp bundle of fur. But by the time the eagle started to circle, he came back to life, realised where he was and what had got him and started to wriggle frantically, spurred on by the calls of his troop below. It was all in vain. The eagle was more powerful and in full control of the situation. It circled once and flew across the river, its triumphant screams drowning the pathetic calls of its victim. We watched in shocked silence as the morbid silhouette shrunk to a speck in the sky. Big Boy had gone for good, whipped from our lives as suddenly and as noisily as he had entered it. The experience had jolted me and, like most profound shocks, it made me see the world in wide-angle. I stared into nothingness and wondered why it was that animals had been made to eat one another; not just to kill but to kill painfully. There was certainly no element of compassion in such a dog-eat-dog set-up. Perhaps a world of vegetarians had been tried and failed. Or maybe the constraints of Nature were narrower than we thought, given the raw materials, and carnivores just had to evolve. Still, I wouldn't have thought it was impossible to have a clause in the rules that said all kills must be painless. To a conscious mind, the world seemed rather callous and filled with uncertainty. I suddenly felt very vulnerable and manipuated.

What I was more prosaically certain of was that giant otters were territorial, but only in a small part – the core – of their home range. In all four family groups, the core contained a cluster of marking sites. Spotted Dick and Co. held exclusive rights over Lama Creek which included Dark Lane and its group of seven marking sites; the Streaky group owned Lama Pond, the south canal between Maduni and Lama, and Bee Canal with its clump of six sites; the Maduni foursome controlled Maduni Creek; and the Anadale group claimed Anira Creek and Anadale Gutter for its own. This gentlemanly arrangement was a stable one; we never once saw families trespass over into neighbouring cores, distances being kept by scent communication rather than by a show of muscle. From our experience, aggressive encounters were non-existent and in all their twenty-odd years of working, hunting and fishing in the area, Coxy and Ram Raj had seen only one angry scene between otters, involving what seemed to be a transient and a resident pair.

The extent of each family group's home range beyond their exclusive core was impossible to measure directly. Not only were the groups difficult to follow but large areas of Russel Lake and the swamps proved totally inaccessible. Radio-telemetry, a useful tool in modern ecology, could in theory have helped tremendously. But it would have meant capturing my otter subjects – an extremely stressful operation and difficult to justify for an endangered species – and fitting them with a transmitter harness. To an otter, a harness can be a very dangerous garment as it is liable to snare and drown the animal. In the open sea off Scotland's west coast, where they have been used successfully on the European otter, there is little to fear in the way of aquatic hooks, but in Guyana, the black waters of swamp and lake abound with all sorts of hidden traps. Without radio-telemetry, it would take far longer to mark out each family group's home range on a map but by making certain assumptions I managed to make a reasonable estimate. It turned out as a minimum of twenty miles of creek or eight square miles of swamp, though by altering the assumptions slightly, it rose to three or four times that amount.

The mere existence of these overlapping ranges was more important than any quibble about their size. And the evidence for them was certainly not wanting. I often spied Spotted Dick and Co. and the Streaky group swimming and fishing – on separate

116

occasions – in the same area in Russel Lake. This piece of common ground lay close to the mango tree cross-over and fell just outside both core properties. The Streaky group also shared the large open area in Russel Lake opposite the Maduni sluice gates with the Maduni family, but again they never used it simultaneously.

The pieces were gradually falling into place. Clearly, territorial defence – exclusive land ownership – did not operate outside the sacred core areas. My giant otter friends had a land-use policy that was rather like the human situation. Human families own a small piece of sacrosanct real estate (their house and garden) but share a much larger area (roads, parks, recreation centres etc) with other families and individuals. The important difference is that by using scent language rather than visual displays or vocal exchanges, giant otters take pains to avoid meeting one another in the common land whereas people don't seem to mind bumping into one another. Sound communication is important only within the close confines of the family group because, in contrast to the likes of wolves and howler monkeys, giant otters have no call that can carry over large distances. This makes inter-group conversations a little less direct but the silent medium of smells – as we knew to our cost – is just as powerful.

11 Games Otters Play

(Keith)

As Nibs got older, keeping him occupied became a full-time job. He was no longer happy simply to swim in the bath by himself, or to chew toes or shoes while we typed or wrote up our camera notes. Those unsophisticated days were past; what Nibs desired was participation. He not only wanted play, his craving now was for play with a companion. Over the weeks he invented or developed a number of participation games, some of them land-based but the majority water-borne diversions. As with most other mammals, his games seemed to be directly related to situations he might have to face in earnest as an adult. The young animal is designed so that what is good for it to practise is also very enjoyable. Nibs' dry-land games were very much linked to attack and defence, two activities that have obvious relevance in the life of any wild animal.

Hide-and-seek was a favourite game: after a tickling session, Nibs would suddenly break free and rush for the end of the bed. This was a signal for the tickler to make off to some other part of the room and hide. I usually disappeared behind the far side of an old cupboard that was set against the bottom of the bed. After a minute or so, Nibs' head would appear warily from under one side of the bed. He would sniff the air suspiciously, then disappear. The head would re-emerge at another site, and the sniffing was repeated until he had reached a point where the bed was closest to my hiding-place.

I was able to watch all this activity without the least chance of being discovered because of a strange otter characteristic. Otters, or at least Nibs, never seemed to look upwards in search of danger or prey. It seemed that anything above a height of four or five feet simply did not exist as far as he was concerned. With any movement around ground level Nibs was far more aware than any human, but happenings above a certain altitude simply did not interest him. This may be a reflection of the otter's way of life; on land the otter is primarily concerned with ground-living prey. No otter can climb, so it would presumably be a waste of time for the species to take much notice of, say, a young bird on a tree outside the otter's reach for the simple and very good reason that the otter has absolutely no chance of catching it. The same bird on the ground is a totally different proposition – it is catchable, and therefore worthy of notice. So it is possible that over the many thousands of years that otters have lived on the earth, they have evolved an internal editor in their brain which stops them being distracted by 'higher' things that they cannot profitably pursue.

A knowledge of this blind spot meant that it was easy to keep track of Nibs' progress along the bed. All one had to do was to poke one's head around the corner of the cupboard while at the same time making sure to keep feet, legs and torso out of sight. As my head was more than five feet above the ground, and therefore non-existent as far as Nibs was concerned, it was possible to watch his careful, infinitely stealthy approach with my head in full view of the little otter but without any risk of being seen. It gave one an inkling of the pleasures of invisibility.

Once he had reached the end of the bed, Nibs would slowly and carefully emerge from its cover, his sinuous body pressed low against the floorboards, tail held straight out behind and a look of deadly concentration on his furry face. When I saw that he had almost reached the cupboard, I would leap out in front of him. Even before my feet had touched the ground Nibs would turn tail and galumph his way back under the bed. As he disappeared beneath the springs, I leapt back into my hiding-place and waited for his next wary advance.

Very occasionally, Nibs would ring the changes. Instead of his slow, careful approach, the straggling sheets would suddenly fly aside as Nibs came tearing out from under the bed, rushing past my hiding-place before I could so much as move and skidding to a rest beneath a low, rickety ottoman. Although there was scarcely more than four inches of headroom under this piece of furniture, Nibs somehow managed to slide effortlessly beneath it and to turn and twist at the same time so that, when he reappeared an instant later, he was lying on his back with just his front paws and legs showing and the occasional hint of a whiskered nose. The idea now was that I was to try to tickle the 'palms' of his forepaws while Nibs defended himself (and very effectively) with his jaws. It was very much a one-sided contest; Nibs' short forelegs were so close to his mouth that I stood almost no chance of successfully completing my side of the game without taking a nip or two on the hand. And Nibs took such fiendish delight in his superiority, insisting on playing the game for up to an hour, that at last I began to cast round for some way of turning the tables. I found it when I noticed the end of his tail sticking out from under the ottoman. I turned my attention to this, giving it a swift tug before rushing my hand to safety. Nibs took longer to reach this source of annoyance and the game became more evenly matched. I even began to enjoy myself. Nibs did not; like all infants, he loved an unfair game where he always came out on top. He crept further under the ottoman and sulked.

I should not like to give the impression that Nibs was always the aggressor in these games. There were times when we imposed upon his good nature, 'wobble-belly' being the most extreme example. Otter cubs are beautiful at any time, but they are probably never so cute as when they have just stuffed themselves full of fish. Then, their long, sleep shape is replaced by a rotund, roly-poly appearance and their fluid natural grace deserts them. After a real blow-out (usually the evening meal), Nibs could hardly even walk – his stomach was so distended it almost dragged on the floor as he staggered along the verandah to his bedroom. At times like these, a young otter wants nothing more than to find a nice quiet corner in which to relax and snooze away the next four or five hours. But on these occasions they are so cute it is almost impossible not to take advantage of them. For one thing, it is the only time an otter is slow enough to stroke and pet. So this was the ideal time for wobble-belly.

I would pounce on Nibs as he made his weary way towards his favourite sleeping corner, swinging him over onto his back. In this position Nibs seemed to be just one big ball of belly-fur from which protruded four short limbs and a long tail. The legs waved ineffectually like a tortoise's and, like a tortoise, it was quite a struggle for him to turn himself upright. But even before he could try I would dive bomb him with my fingers, seizing his bulbous belly and shaking it back and forth. At such times Nibs presented a truly comic picture – his limbs and tail all stayed perfectly immobile while his titanic tum wobbled to and fro, seemingly independent of the rest of his body. I'm not sure that Nibs actually enjoyed wobble-belly – he was usually too overfed and sleepy either to fight back or to take much pleasure in the game – but for Liz and myself it was the high point of the evening.

The wobble-belly game had another advantage: it let one play with Nibs in a seated position. 'The chase' did not. Nibs invented this game when he was five months old. We had by this time evolved a strategy for escaping from his room with the minimum of fuss. I saved the largest fish until last, then threw it into the farthest corner of the room. The theory was that Nibs would rush for the fish and by the time he had consumed it and realised he was alone, we would be happily enjoying our siesta.

I must have misjudged the amount of food I had already fed him because when I tossed the half-grown piranha across the room and turned to flee, Nibs followed me and not the fish. Liz was waiting at the door, having left a few moments earlier, ready to swing it shut the instant I was through. Unfortunately, Nibs was so close on my heels that Liz couldn't shut the door without cutting our otter-baby in two, with the result that Nibs went scampering along the verandah with two very irate humans in his wake. The chase continued around the house, with Nibs on several occasions going perilously close to the edge of the verandah with its fifteen-foot drop to the rock-hard ground beneath. Throughout it all we gave voice to various endearments in an attempt to bring our wandering 'son' back to the bosom of his family. The calls ranged from 'Come on, Nibs baby' at the beginning of the chase through 'Stop, you little bugger' to . . . well, to words that are better imagined than written.

Suddenly Nibs did stop. He spun round and faced us, his front paws extended, his head low and his tail swinging gently from side to side. It was obvious that Nibs expected us to turn tail and run while he took his turn at chasing us! At first I thought of pouncing on him there and then, but it suddenly occurred to me that we could use Nibs' request to our own advantage. Liz and I quickly fled back along the verandah with Nibs in hot pursuit. I reached the door of Nibs' room first and, swinging it open, hid myself from view. Liz ran on into the room and Nibs in the excitement of the chase followed her through the door. Liz had meanwhile stepped to one side of the door and when Nibs ran through she leaped back out. I shut the door and Nibs 'wheeped' his disappointment, crying piteously through the mosquito netting that covered the lower half of the door.

If we had got the better of our young otter on that occasion, Nibs was perfectly capable of using the chase for his own ends. We would be in full cry after him as he ran along the verandah when, without warning, he would make an abrupt right turn and we would find ourselves in the bathroom, with Nibs already in the large shower area, waiting expectantly in the six inches of water that we always kept in there so that he could have a cooling dip whenever he wished. But Nibs did not feel like a refreshing swim; from the way he sat tense and expectant in one corner it was obvious that he expected the chase game to continue inside the shower! As an otter, Nibs simply would not believe that anyone could prefer dry land to water. So, not wishing to hurt his feelings, one of us would jump in with him and spend the next half-hour leaping about like a mad thing, rushing around the

121

perimeter of the shower and playing in turn the part of pursuer and pursued.

It was a delight to watch him: there was a definite, joyous play-signal in his movements. Chimps and orangs, and humans too, make a 'play-face', an open-mouthed, half-smiling expression when engaged in games. The otter can make no great change in its facial expression, but it can, and does, change its gait. On land, Nibs galumphed along the floor to show he was playing, bouncing his way about the room in a perfect parody of an adult otter's gallop. In the bath this galumphing became even more pronounced, with Nibs bounding around the bath in a series of lamb-like leaps.

Sometimes Nibs would leave the water and continue the chase on terra firma once again, but more often than not he seemed only too happy to gambol about in the shower forever. In an attempt to lure him out of the water we would jump from the shower uttering hideous cries of fear, in the fond hope that Nibs would follow. This worked for the first two or three times, but Nibs got wise to the ploy. Then he simply stayed put in the farthest corner of the shower, sinking his head low between his forepaws and wriggling his tail every time we appeared. The unspoken message was easy to understand: 'I still want to play, but in the water please.'

If we decided we did not want to play, Nibs was perfectly capable of devising his own aquatic amusements. When the mood took him he could play alone for hours in the shower with a collection of weird toys that gradually increased in number. By the end of the trip Nibs had acquired a piranha jaw, an old stick, a camera brush, two hollow plastic balls, a plastic film-case and a mini-coconut. One toy that we did not like Nibs playing with was the plug of the shower. Nibs had what amounted to an obsession, a fetish, over the plug. He would fight for hours to pull it from its secret socket deep in the floor of the shower. At first this was an easy job: he simply slipped his upper canine tooth into the attached metal ring, pulled, and up it came. Things became more difficult when I removed the ring, but Nibs soon learned to insert his eye-tooth into the rubber recess which had formerly taken the metal ring. When I cut this off too, making the plug completely flat and fitting it flush with the smooth tiles of the bath, I truly believed I had Nibs beaten. Not so; it only took a little longer to dislodge. We learned to dread the sound of water rushing from the shower pipe into the creek. It meant that Nibs had once again solved his self-imposed problem and our hour of peace had ended.

122

Within a minute of ending this game Nibs would realise that the essence of all his pleasure – his beloved bathwater – had mysteriously vanished! Then, to remedy the situation, he began to wheep. We soon found that the quickest way to stop the noise was to refit the plug (I took to hammering it into the hole) and refill the shower. But Nibs never did make the connection between plug-removal and the loss of his precious bathwater. As soon as he was re-admitted he made unhesitatingly for the plug!

Whenever Liz played with Nibs she invariably ended up in a game of 'figure-of-eight'. This was partly because Nibs loved the game and partly because it helped to avoid a lot of painful biting. When Nibs had leaped upon an offending hand and was about to give it the dental treatment, the attack could be stopped immediately if he was flipped onto his back and pushed and pulled through the water. This was especially effective if a thumb was placed in Nibs' mouth and he was allowed to wind his forepaws around the wrist of his playmate while the action was going on. In this position Nibs went completely immobile for as long as he was towed about the water in circles, straight lines and figures of eight. His long body went completely flaccid and his little dark eyes seemed to take on the staring, unfocussed look of a hypnotic subject. Nibs never bit the thumb in his mouth during figure-of-eight, he used it simply as an extra handle to prevent him losing his grip. As with the metronome game, rhythmic movement seemed to calm Nibs, almost to mesmerise him. While he was kept moving he was happy but stop for a moment and he showed his resentment by reverting to his normal otter self, slipping your digit into the side of his mouth and chewing hard with his needle-like molars!

Apart from figure-of-eight, most of Nibs' other games were quite obviously related to improving his aquatic hunting skills. He loved to chase after a spare piece of nylon rope with a large knot in one end. We had only to swirl it through the water in front of him and the hunt would be on. No matter how hard we tried, it was impossible to keep the knot away from his snapping jaws for more than a minute at a time. We marvelled at his aquatic skill, his ability to dive instantly and change direction, to turn on a sixpence in pursuit of his elusive nylon quarry. Few fishes would escape Nibs once he had them in his sights, and we felt a little more confident about his chances of survival when he was finally old enough to be returned to a more natural existence.

A similar hunting game evolved when Liz threw a new toy, an old plastic film-case, into the water for Nibs to play with. Instead of the usual pounce and bite, release, pounce and bite cycle that the young otter normally indulged in, Nibs grabbed the case, bit it lustily and then brought it back to Liz. He even allowed her to take it from his mouth, an unheard-of liberty as such attempts had previously resulted in serious attempts to bite. On this occasion Liz sweetened the robbery with a twenty-second tickling session before once again throwing the case to the far side of the shower. That set the pattern. Nibs immediately retrieved the object, accepted his tickle and rushed off to intercept the next throw. When he was tired of this Nibs turned the film-case into a surrogate fish, pushing it with his nose to make it swim in front of him in imitation of live prey. He seemed to crave such activity, and it is very likely that this hunting behaviour was wired in to Nibs' brain, much as stalking behaviour is with cats. The German ethologist, Iranaeus Eibl-Eibesfeldt, has described how cats literally cannot do without a daily period of stalking. Even when kept in a completely bare room with no playthings, the cats go through the whole sequence of careful stalking followed by the characteristic short rush and the final, killing bite.

On the rare occasions that his toys were missing, we discovered that Nibs was capable of procuring new ones. I looked up from my camera notes one day to see a furry brown body making its surreptitious way towards the bathroom with our can-opener in its mouth. Following quietly, I watched Nibs carry his prize to the shower, climb in and release the iron fish into the water. A look of astonished disappointment spread across his whiskered features when he discovered that iron fish cannot swim. After much pulling and poking in an attempt to jolly his plaything into movement or flotation, he gave up, took it once again in his jaws and made to leave the water. He had just got out when he noticed that I was watching. With a presence of mind that any veteran shoplifter would have admired, Nibs wriggled swiftly under an old carpet we kept next to the shower, and emerged from the other end minus the can-opener! He then marched past me with an air of perfect innocence and began the search for another, more buoyant toy.

Nibs enjoyed all these inanimate toys but there was no doubt that his especial joy was hunting the live fish that we put into the shower. This may sound cruel, but otters have a strange way of altering your moral outlook on life. He so obviously enjoyed rushing after the fish in his bath, revelling in the pursuit and capture with a totally un-self-conscious glee, that I found I could cheerfully throw in four wriggling patois and watch them being eaten alive with nary a pricking of my conscience. Under Nibs' influence, it all seemed so natural.

124

Nibs was fed most times with fish we had caught ourselves. These were usually red piranha and patwa, both of which could be taken in abundance with a number 6 hook and a little bread paste. We were amazed by the ease with which Nibs caught them one after the other when we fed them to him live. I had seen patois shoot away at high speed when they caught sight of me on the bank, and the force with which the red piranha hit my line spoke of great powers of acceleration. Despite this, Nibs seemed to take the fish effortlessly. It was true that they did not have a large area in which to make their escape, but even so his performance was remarkable. At times, the fish made only half-hearted efforts to escape, so that I began to feel that Nibs had some strange power over them, that somehow they knew that his presence spelled inevitable doom. Nibs often had to resort to chivvying them along from behind before they would put on a burst of speed worthy of an otter's attention.

Once he had caught his chosen fish, Nibs seldom ate it immediately. He had an almost contemptuous disregard for their powers of escape and many times he would hold a fish underwater with one paw, sweeping it around the bathroom tiles like a floor-cloth, while at the same time casting about him for more difficult prey to pit his skills against. Having tried (and usually failed) to grasp squirming, slimy, freshly-caught patwa on the creek bank, I was filled with admiration at the absent-minded ease with which Nibs held his catch fast.

Nibs' fishing technique was also improving outside, in the waterways around the rest-house. We did not yet dare to release him completely, so he swam attached to a long nylon leash some fifteen feet in length. Despite being perfectly at home in the water of his shower, Nibs showed a great deal of initial distrust on his first meeting with a real creek. The small canal he had swum in at Cane Grove had not prepared him for the big wide river and he approached this novel body of water hesitantly, his trunk flattened hard against the grass bank, his head and tail down, and ready for instant withdrawal. And yet, despite his fear, there was something about the creek that resonated with Nibs' ottery brain and drew him ever closer to the edge of the bank. At the water's edge he stopped, sniffing the still surface. Then his nose went in and almost immediately came out again as he retreated a few paces; he stopped, looking up at me as if for reassurance and then advanced again, pushing his head in to the level of his ears this time. I could see that he had his eyes open all the while, though at first he did not seem impressed by what he saw as he shuffled back a foot or two once more as if he had seen quite enough for one day. Then suddenly, in one swift, fluid motion, he had poured himself over the side of the creek and was swimming and diving with a sort of mad passion that we had never witnessed while he was in the shower.

This was an aspect of Nibs' personality that we were to see more often, whenever we took him into river or lake. At these times Nibs became imbued with a frantic zest for life: his actions were faster, his reflexes sharper and the muscles along the whole length of his sinuous body took on a different, tenser tone. His senses were so acute in this mood that he seemed able to hear even faint noises underwater. On one occasion he had dived deeply when a spurwing called on our right, far inside Russel Lake. Nibs surfaced two or three seconds later and immediately looked in the direction of the sound. We began to realise that, cute as Nibs was, and happy as he appeared to be at Lama, there was really no other way for an otter to live than completely wild and free. Only then did the true essence of this most beautiful of all the carnivores reveal itself completely: a captive otter was really no otter at all.

But before we could release Nibs we had to be sure that he could fend for himself, and these leash-directed swims were a first step in introducing him to the outside world in which he would eventually have to make his way. The lead prevented Nibs from leaving home too early and allowed us to protect him from the ever-present danger of predators. Anaconda and caiman were common around Lama and we always kept a cutlass to hand should the worst happen.

On one occasion the worst did happen. After a long swimming session, Liz was leading Nibs by his leash back to the rest-house when he unexpectedly pulled to one side, dragging the leash from her hand. In a moment he had dived from the bank and was swimming strongly towards the reed beds on the far side of the canal. We screamed at him to return, but it was as if Nibs had suddenly become deaf. He swam steadily on, diving occasionally, and finally climbed out to rest on a flat patch of dying reeds. It was then that we saw it: about thirty yards from Nibs was a long, dark shape. At first we thought it was a floating log, but then we realised with horror that it was moving slowly against the wind. A caiman! And it was making straight for Nibs.

We yelled increasingly desperate warnings to our still-deaf otter, calling impotently across the water as the caiman slowly narrowed the gap between it and Nibs. I looked down the water-way at the rest-house, judging the distances. There was no hope there, no chance of running to the moored canoes and returning in one of them before the caiman had closed with Nibs. What could we do? The caiman got steadily closer but Nibs was still blithely unaware of the approaching danger. I looked round for something to throw at the reptile, hoping to scare it off, but there was nothing to hand. I reached for the cutlass. I would only have one chance, but perhaps a lucky throw would discourage the beast. At that moment Liz spun round. 'Do your "tch" sound! Quickly!' she urged. I saw her plan immediately. Whenever I fed Nibs I used a 'tch-tch-tch' call and he eventually came to link the sound with food. As Nibs was perpetually hungry at this age, at the first sound of a 'tch' he would come running to my side, rising on his hind legs and sniffing the air expectantly for fish. We had used this ploy several times when Nibs had wriggled his way into some inextricably tight corner under the rest-house furniture.

There could be no tighter corner than this, I thought despairingly, and I set to with a will. The problem was that the 'tch' sound is a sort of whisper, and it is virtually impossible to whisper loudly, especially when the sound has to carry across sixty feet of water and the recipient of the message is playing at being hard of hearing. The caiman was by now not more than twenty feet from Nibs, and I 'tch-tched' frantically, with plummeting hopes of saving our little otter. But I had forgotten Nibs' love of food and his super-awareness when in open water. After a short pause he lifted up his head as the 'fish-fish' sound drifted across the water. Then, with the caiman only fifteen feet away, he sprang from his resting-place into the water. For an instant I thought his nylon leash would wrap around the reeds and prevent his escape, but the rope glided smoothly between the stems and soon Nibs was halfway across the canal. The caiman immediately changed course and tried to follow its rapidly disappearing lunch, but it was no match for a speeding, hungry otter. A moment later we were pulling Nibs from the water. He was still quite oblivious to the doom that had almost overtaken him, and was far more interested in searching my pockets for the non-existent fish my calls had promised.

When we told Katina, the watchman, about our adventure he gave little sign of interest. But the following evening he left the rest-house by canoe with his shotgun across his knees. Later that night there was a single gunshot, and the next morning we found a dead caiman in Katina's canoe, its head blasted by shot. At first we believed that the watchman had destroyed the caiman because of the threat it posed to Nibs. We were correspondingly grateful for the trouble he had taken, presenting him with three packets of cigarettes from the store we kept for just such acts of kindness. We also persuaded him to pose with his shotgun and his trophy

127

for a photograph. It was only later that we discovered that Katina had a seine net close to the rest-house and had killed the caiman out of pure self-interest. Left alive, the reptile would undoubtedly have raided the net for fish, destroying much of its fabric in the process. On close inspection, the caiman proved to be lacking its left forepaw, though the wound had been fully healed before it was shot. 'Pirai', Katina explained laconically, meaning that a lone piranha had probably attacked the caiman earlier that year, removing the paw with a single bite. That was another danger that Nibs would have to face when he was eventually released, and both incidents – the caiman and the piranha – brought home to us just how hazardous these quiet-seeming waters really were.

Although times of crisis are the most easily remembered, the majority of Nibs' days in the creeks were spent in carefree play. As in the shower, he loved to retrieve objects, only at these times his toys were anything that came to hand: a stick, leaf or – his outdoor favourite – the red, pear-shaped fruit of the cashew tree. It was wonderful to see him attacking these objects from below, zooming in like a fighter pilot on his foe, his eyes wide open and his mouth snapping shut at just the right moment to seize his prey. But on many occasions Nibs would forsake the pleasures of the open water and probe and pry his way between the gaps and crevices of the old jetty; he seemed irresistibly drawn to these dark underwater caverns. On one occasion there was a flurry of activity under the planks and we thought at first that Nibs had disturbed an anaconda. But, after a moment's pause, we heard the unmistakeable sound of an otter eating fish and realised that Nibs must have caught his first wild prey by trapping it in the confined space under the jetty. It was another indication that Nibs was growing up.

Nibs' crevice-fishing technique also gave us a clue to the problem of co-existence between the giant otter and individuals of Nibs' race. Scientists had long been puzzled by the fact that both the Guyanese and the giant otter were to be found on the same stretches of river. Why did they not compete for food? Nibs' behaviour seemed to indicate that at least part of the answer was that the two species' hunting habits were different. The giant otters, we knew, hunted either by deep diving or by splashing their way through shallow water in the hope of disturbing lurking fish. It seemed that the smaller Guyanese otter spent a lot of its time investigating crevices, trapping its prey there. It was likely that the fish caught in this manner were of different species from those taken by the giant otter or so small as to be useless to the larger species. In effect, the otters had divided up the habitat

128

between them and as long as neither encroached on the other's patch, they could happily co-exist.

Nibs' discovery of his natural element made him less of a friend to Mutt and Jeff, the two half-grown mongrel pups that lived in and around the rest-house. Although quite aggressive towards Nibs when he first arrived, the two dogs soon learned that he was a permanent resident. Gradually he came to be accepted and allowed to join in their games of wrestling and catch-as-catch-can. All this stopped as a result of Nibs' love affair with the water. Try as he might to persuade them, the two dogs simply would not agree to go swimming with him and so the warmth of their friendship cooled perceptibly. But the two dogs were still inseparable and after losing Nibs, they took to playing by themselves on a small patch of clear ground near the waterside, midway between the jetty and the jungle.

About two weeks after Nibs' first dip in the outside waters, Mutt came screaming into the rest-house, squealing and barking piteously. Something was obviously very wrong and we followed the whining puppy back along the canal bank to the dogs' favourite playing spot. There were signs of disturbance everywhere: beaten-down vegetation, broken sticks and signs of scratches in the bare dirt. Most ominous of all, there was no sign of Jeff. After a hurried, fruitless search we ran off to Katina, the nominal owner of the two dogs. He came reluctantly to the spot and, having viewed the signs, diagnosed an attack by an anaconda or 'boa constrictor', pointing out the gap in the vegetation where the beast had fled.

The watchman did not seem too bothered by Jeff's disappearance. Snakes are just a fact of life in Guyana, an unavoidable nuisance, like aircraft noise in England. And besides, as Katina said, his bitch could always produce more puppies, so why worry? But we could not be so complacent. To be honest, we were terrified. What, we thought, would happen if the boa constrictor got peckish again and decided that otters were quite as tasty as dogs? No, the boa would have to be removed. Katina showed no interest in helping us to catch the reptile, so we made a half-day journey to Maduni Creek to visit Iburu, an Amerindian with a famous reputation as a hunter. He was a squat little man with a squashed-looking face and an enormous belly, but his unprepossessing appearance belied an extensive knowledge of the natural world. Whenever a hunter wanted a guide for this part of Guyana, his first choice was Iburu. After a little haggling over his fee, Iburu agreed to accompany us back to the rest-house and to catch the boa there and then.

Almost as soon as he stepped ashore, the Amerindian scotched Katina's theory that the snake was a boa. From the wealth of signs on the ground, quite meaningless to us, Iburu deduced that the attack had come from the water. The assailant was without doubt a 'camoodie', the Guyanese name for the anaconda. According to Iburu, the snake likes to sneak close to its victim along a waterway before suddenly darting forward, seizing the prey with its jaws and throwing loop after loop of its long, muscular body around the quarry. If it is too far from the water it will squeeze the victim, asphyxiating it. But if the anaconda is close to water and the slope of the land favours it, the reptile does not bother to constrict its victim, it simply rolls swiftly to the water's edge and, still holding the prey in its coils, falls in and sinks. A reptile can live for far longer under water than any mammal, and by this strategy the snake is able to kill its victim without wasting all that valuable energy in squeezing it to death. Iburu claimed that this was what had happened to Jeff and that by now the anaconda should have surfaced again to swallow its prey. With that, the hunter stalked off slowly along the bank, sniffing the air delicately. We followed at a discreet distance until after about a hundred yards Iburu stopped and beckoned us over.

He sniffed again. 'Smell,' he whispered. We did, but could smell nothing. 'Smell!' Iburu repeated imperiously, as if angered by our effete, townsmen's senses. We complied and there, just for an instant, I was aware of a stale, rancid odour. Liz smelt it too, wrinkling her nose in disgust. Iburu saw her expression and nodded, pleased at our success. 'There, there,' he breathed, pointing under an old Ite palm stump that lay lodged against the river bank. 'Camoodie there.' I leaned forward and could just make out the squat, ugly head of what looked like a large anaconda. Iburu went back to his canoe and returned with a long bamboo rod with a wire noose at one end. With infinite slowness he manoeuvred the noose closer and closer to the dark, unmoving head of the reptile. At any instant I expected the snake to draw its head back into the burrow it was occupying, but the hunter's movements were so exquisitely slow that the anaconda seemed totally to disregard the presence of the noose, even when Iburu slipped it under its jaw and worked it back behind the head.

Then Iburu struck. It took enormous strength to jerk the heavy bamboo staff up with sufficient speed and force to trap the snake, but Iburu was a master of his trade. The noose was pulled with such violence that I could see the wire biting into the snake's flesh. The snake came suddenly to life, trying vainly to escape, pulling its head into the riverside cavern, but succeeded only in tightening the noose still further around its neck. Iburu passed the bamboo to me, warning me to hold it tightly while he went down in the water to complete the capture. He took with him a second bamboo about fourteen feet long and a cutlass. Once in the water, Iburu quickly closed with the snake which tried to withdraw into its hole with such force that the bamboo pole was almost wrenched from my hands. Iburu pushed the stick into the fleshy mass of the snake's body, and as the reptile tried to coil around it, the hunter's machete whirled in a silver arc and the noose suddenly came free, bringing with it the severed head of the anaconda!

We were completely astonished by this unexpected turn of events. We had planned on catching the snake alive and sending it to Martin, the Snake Man. Iburu, having spent most of his life with hunters, had completely misunderstood our requests. I stood in shocked silence for a moment, on the brink of giving Iburu a dressing-down for his hasty actions. But it was too late, the snake was dead, and we needed Iburu to guide us in many of the more remote areas around Lama. So I kept my cool and merely asked Iburu to bring the rest of the body to the shore. He did, and we discovered that the snake measured all of twelve feet without the head. In total it was close on thirteen feet long, and halfway down its sinuous body was an ominous lump. We cut the snake open and in its stomach found Jeff's body, already half-digested and almost unrecognisable except for the distinctive sandy colour of the fur. It was yet another reminder of the dangers that lay in wait for Nibs as soon as he said goodbye to Lama and the protection of his human parents.

12 The Rains

(Keith)

The lead-up to the rains brought a change of pace at Lama Creek. The heat of the dry season had forced us to do everything slowly. Now, just before the five-month rains, we hardly moved at all. The weather was unbelievably muggy, the air so full of water that its weight pressed down on one's head and shoulders, making it difficult even to think. The atmosphere was stifling; it was like breathing wet flannel. A swim in the cool, dark waters of the creek was almost irresistible, but the thought of piranha lurking just beneath the surface deterred us. We showered four, five times a day but felt no better. Pappy and Katina solved the problem by lying in long-armed berbice chairs and staring at the ceiling for most of the day.

Even the animals seemed to feel the overpowering stillness of the days, especially the otters. Nibs was unusually quiet and, when we forced ourselves into the canoe and out onto the creek, both the Streaky group and Spotted Dick and Co. were absent too. The only sign of life was the buzz of dragonflies as they patrolled the waters in search of prey, and the occasional glimpse of a capuchin troop. The monkeys seemed to be emulating Pappy and Katina; they hardly moved at our approach but simply lay where they were, draped limply across the tree branches in a variety of poses and hardly able to turn their heads to follow our slow progress down the creek. Once we came upon three hoatzin perched low in the sweltering gloom of a banza tree. The birds were panting madly, their only method of temperature regulation. On the other side of the dyke, in Russel Lake, nothing moved. Without wind, even the dry, metallic rustle of the Ite palms was silenced. The reed beds shimmered as the water level fell steadily, bringing supply problems to Georgetown. There were rumours that taps in the capital were producing worms as well as water! Everyone was waiting for the rains.

They finally broke late one morning, just as Liz and I had finished exploring a small tributary of Maduni Creek. We had spent the night at Iburu's hut, sleeping under a palm-thatched

132

lean-to in large Wapisiana-Indian hammocks, about three times as wide as the naval variety and far more comfortable. It was a lot cooler in the open than in our room at Lama, but fears of mosquitoes and vampire bats made us cover the hammocks with mosquito nets, an operation that required two dozen safety-pins and no small amount of ingenuity to block every point of access. The night had been enlivened by a squadron of bats – insect-eaters not vampires – who had suddenly decided that the air space under our lean-to was the ideal place to hunt. Our sleep had been punctuated by the sweet sound of bat after bat crashing into the mosquito net which was apparently too thin for their sonic radar to detect. I noticed that Iburu's three dogs, who had stationed themselves beneath our hammocks, greeted every new catch with snarls and growls. Their noisy bustling was disturbing but I did not think too much about it as the passage of the bats was quite enough to occupy our attentions. They fell in fits and starts down the nylon meshwork, chittering angrily as they passed our heads, and crashed noisily to the ground.

At least, most of them did. I awoke in the middle of the night with something fluttering insistently at my face. I swatted it away absent-mindedly, then suddenly sat bolt upright in the hammock. Somehow, a bat had slipped in under the protective net and was now flying in panic-stricken circles around the inside of the nylon tent, desperate to escape. I was equally horror-struck. The touch of those leathery wings on my face had awoken in me a primordial fear. I had to get out! With trembling fingers I fumbled at one of the many safety-pins. It opened and I dived head-first out of the hammock, landing slap in the middle of a sticky, furry pile of I-couldn't-see-what. My fearful gibberings had woken Liz by this time and she shone her pen torch in my direction. It was then that I discovered I was lying in a pile of chewed-up bat corpses! Iburu's dogs had been doing more than just bark at the bats: each time one had fallen to the ground they had leaped upon the unfortunate creature and bitten it to death. The smell and feel of their mangled bodies was more than I could stand. Feeling like a recipe for a witch's brew – wing of bat and various other revolting ingredients covered my clothes and skin – I made straight for the river. For once I ignored the piranha and caiman and leapt in, stripping off my clothes in the water and scrubbing my skin obsessively long after the bat remains had been removed. Then, accompanied by a train of eager mosquitoes, it was back to the hammock. It had not been the best of nights.

Next day we checked the narrow, winding tributary of Maduni Creek, searching the banks for recent spraint. The lack of sleep,

133

the overpowering humidity and the fact that we found absolutely no sign of otters after five hours of hauling our canoe over fallen logs and shallow water left us feeling pretty miserable as we pulled out of a claustrophobic side channel into the main creek. As we paddled, we were suddenly aware of a dull drumming in the trees, a noise that was getting louder by the second. We looked in the direction of the sound and for the first time noticed that, instead of the usual cloudless blue we had become used to, the sky to the west was now an angry, inky black. It was an eerie experience, for the sun was still shining through a crack in the looming black mass. As we watched, around the bend in the river came what I can only describe as a golden waterfall. It was the first real rain for months, back-lit by the orange-red light of the sun, and it was moving up the river towards us like a golden cascade of liquid light. We watched spellbound as the honey-coloured rods of water approached with ever-increasing speed, then we were inside the cloudburst and the first drops splashed tingling cold on our overheated faces. We gloried in the coolness, and it was not until we were drenched through that we realised our problems. Golden it might be, but this enchanting spectacle was still rain and it was still very wet. Our cameras could not appreciate the artistic beauty of the storm! Hurriedly, we closed our camera cases down tight, making a mental note to put more silica gel and uncooked rice (both of which absorb water) around the wetter items. Then once again we gave ourselves up to the icy delights of our first cold shower for months.

The animals, too, responded to the coming of the wet season. The capuchins and the howlers seemed to find the rain even more tiresome than the pre-rain mugginess. They simply disappeared. Whether they were sheltering from the rain in some inaccessible spot or had moved on to greener tree-tops we were not sure, but there was no sign of either troop. On the other hand, animals that were usually secretive became more open in their activity. This was especially true of the caiman. Apart from the incident with Nibs, we had seen caiman only rarely. Now they seemed to be all over the place. They were particularly common at night, and Liz and I would often spend half an hour with a torch scanning the creek waters. The eyes of caiman reflect torchlight with a golden-white intensity; they never seem to blink, and it is quite eerie to watch an unwavering gaze staring back at you from the darkness. Pappy had at one time hunted caiman and he showed us his skill at calling the beasts. Pursing his lips, he sucked air through them in short bursts, making a high-pitched, squeaking sound. After a few minutes we noticed a white swirl of water on the surface of the creek and, shining the torch, found a large caiman only eight feet from the jetty, with two more about six feet behind the first. Pappy told us that the noise he made imitated the sound baby caiman produce when they first dig their way out of the underground nurseries in which the female caiman lays her eggs. The infant's calls bring the mother over to the nest to care for the still vulnerable young. But even without Pappy's calling, the creek seemed full of caiman. It was a bad night when we did not spot seven or eight caiman during our torchlit sweep of the creek.

Other reptiles were more of an immediate problem. On the fourth night of the rains I got up around midnight and plodded my way down to our primitive loo. Getting up after sitting there a while, I caught sight of a long, scaly object which slipped out from under the seat and hung down groggily inside the bowl. My legs turned to jelly as I realised what it was – the head of a labaria, a fer-de-lance! It had been coiled round the top of the bowl, between the porcelain and the wooden seat, and the only thing that had prevented something very nasty happening was the fact that my weight, as I sat meditating, had pinioned the snake between the two halves of the loo until I stood up again. The snake was groggy from its crushing (though not completely squashed because of the rubber stops on the bottom of the seat) but it was rapidly returning to its normal, angry self. Without even waiting to pull up my pyjamas, I turned and fled. Or rather, I tried to. With my trousers still around my ankles, I succeeded only in falling heavily to the ground. But at least I fell in the right direction – through the door. Scrabbling madly from the loo, I slammed the door shut, at the same time calling on Liz to come downstairs and to bring a cutlass with her.

The commotion brought Pappy from his bed, a prospecting knife in one hand and a hurricane lamp in the other. He made no attempt to protect my dignity as I gabbled out the story, but simply doubled up with laughter before I was even halfway through. I got angrier and angrier, and finally declared that I was going back in there with a cutlass 'to finish off the bloody animal'.

'Nah, man,' Pappy gulped out between giggles, 'yo' aksin' fuh trouble.' He held up the hurricane lamp. 'We got de better way.'

I thought at first that Pappy meant the lamp would make it easier to see when it came to chopping the snake. But to my surprise he snuffed out the light and proceeded to pour kerosene into a large mug he had asked me to fetch from the kitchen. When it was full, he positioned himself in front of the loo door with a torch in one hand and the mug in the other. Tensely, he gestured that I should open the door, but very slowly. I lifted the latch and gingerly pulled it open. The dark cavern of the loo gaped menacingly, and at any moment I expected the slightly squashed shape of a fer-de-lance to come slithering out, bent on revenge. Pappy waited in the darkness until he could see all of the loo. Then he switched on the torch. He played it up and down the small room, then held it steady. There in the far corner, half hidden under the U-bend, lay the coiled form of the fer-de-lance. The snake seemed to have recovered completely from its recent ironing. Its head was raised threateningly and its forked tongue flicked in and out, scenting for danger.

Pappy had been standing immobile, but now he suddenly came to life. He took a step forward and threw the paraffin over the snake in one rapid movement. The fer-de-lance puffed itself up aggressively at this attack, but to everyone's relief it stayed where it was. Pappy followed up immediately. He drew a box of matches from his pocket, lit five matches together and cast them straight at the snake. In the darkness, he seemed like a miniature Zeus firing thunderbolts at his foe. The snake, thoroughly doused with paraffin, erupted into flame as soon as the matches struck home. It tried to make off, but the flames were too powerful. It was like a scene out of hell as we watched it in the centre of the fire, writhing on the floor while the flames consumed it alive. Soon it was no more than blackened husk.

It was horrible, but I had to admit that Pappy's ruse was by far the safest way to deal with so deadly a snake. There was a risk in throwing paraffin over a wooden floor and setting a match to it, but that danger was far preferable to facing the fer-de-lance with a cutlass. With Pappy's method there was little chance of getting bitten. At no time had he ventured closer than five feet to the reptile. I filed that bit of information away for future herpetological crises.

Nibs' reaction to the rains was one of unalloyed joy. He revelled in each downpour, and the heavier it was the better he liked it. One of his favourite haunts at this time was below a broken gutter where rainwater that had collected from the rest-house roof fell like a miniature waterfall onto the bare ground. Here Nibs would play for as long as the storm lasted. He loved to run through this overflow, only to turn and run back again and again. At times, he seemed to think the waterfall was solid and he would leap again and again at the seemingly concrete spout of water, trying vainly to bite it. Another ploy was to lie on his back with his head right in the rushing stream of water, his jaws opened wide to let the water play on the roof of his mouth. We had seen the same sort of activity some weeks before when Nibs interrupted Liz's wash-day by leaping bodily into a sink full of soap suds and dirty clothes. Each time the water was turned on, Nibs dived under the ensuing stream of water to wash the roof of his mouth.

Nibs' trampling under the broken gutter gradually produced a mud-hole in the bare earth, and a few minutes of play would convert our little otter from a cuddly ball of fur into a miniature version of the Creature from the Black Lagoon. Thus accoutred, Nibs seemed to take a perverse delight in baiting Pappy. He would wait until the cook had finished mopping out the rest-house floor and then rush from his mud-hole onto the pristine floorboards, wriggling about on his back and leaving muddy hieroglyphics in his wake. Pappy would explode, screaming curses in Hindi at the otter and throwing his hands about in a perfect imitation of an irate Italian. Nibs would look up innocently as Pappy stormed into the kitchen, declaring that 'either that devil-bred wata dog go, or I goin' fuh go'.

Despite his outbursts, Pappy was an animal-lover at heart, and we relied on his good nature whenever Nibs' fish supply ran short. Our cook was the finest spear fisherman in Demerara, and he could always be prevailed upon to pit his skill against Lama Creek's fish. On our first spear-fishing trip together we saw several large hurri – a fish which normally lurks hidden on the bottom of the creek – bobbing placidly an inch or two below the surface.

Such fish were easy game, but the most prized of all fish, the lukanani, rarely presented so simple a target. Usually, the only sign of its presence was a dim, black-and-gold shadow floating about three or four feet underwater. Here great skill was needed. Pappy had never had a physics lesson, but he knew all about refraction and the way that light travelling from water to air bends so that an object underwater appears in a different location to its actual position. Pappy could judge this to a nicety, and we never saw him miss a cast with his three-spined spear. Once, however, he did slip as he threw the trident, and the spear did not completely transfix the fish. With great presence of mind, Pappy kept his weight on the spear, holding the fish to the bottom. He called me over and asked me to keep the lukanani pressed against the bed of the creek while he brought round a canoe and secured Nibs' supper. The lukanani was not exactly keen on this idea; it wriggled desperately and moved slightly along the creek floor. I shifted my position to keep above it and stood, unknowingly, on top of a small fire ants' nest.

I had not noticed the nest during the day, but fire ants are particularly active before and during the rains when they breed. The ants had begun to forage in our rooms several days before (possibly drawn to the smell of rotting fish carcasses Liz kept there for reference). They were a complete nuisance, swarming over our bed during the early part of the night. Although less than 4 mm long, the fire ant is possessed of an enormously potent, fiercely burning venom.

I discovered this as soon as I planted my feet atop the ant nest. A colony can contain close on half a million individuals, all unbelievably bellicose and willing to take on any creature they feel is threatening their domain, Man included. My ant nest was much smaller, but that didn't seem to cow its inhabitants. My feet and calves felt as if they had been plunged into molten metal. I tried to withstand the red-hot agonising pain for as long as I could (the fish was big and would feed Nibs for a day or more) but within thirty seconds I had dropped the spear and was running, cursing and yelling, from the site.

After scraping the ants from my legs I found that I was covered in small red weals. The pain slowly receded over a period of about two hours. But as it lessened, it was replaced by an almost unbearable itchiness that was hardly soothed by the anti-histamine creams and lotions that we carried in our medical kit. A few hours later each sting had become a small blister. These gradually turned purulent and the bites were still visible a week later. Some eventually left scars.

I collected other reminders of life in the wet and wonderful tropics but fortunately they were rather less painful. I filmed downpours, animals in downpours and Liz working in downpours. The camera equipment did not like the damp conditions and despite being sheltered, often to the detriment of the photographer, it still absorbed water from the humid atmosphere. In conditions like these no amount of rice, rubbing or blowing with my air-gun could prevent the inevitable: the first minuscule signs of fungus on the lenses. It was all I could do to prevent this ogre from spreading its web-like tentacles all over the lens. Even after the rains had broken, the air was still extremely humid, if a trifle breezier. Material I had never dreamed of as absorbent acquired a sponge-like quality; paper went limp with dampness, becoming difficult to write on, books curved, toothpaste ran and our instant coffee coagulated into a sticky lump. We had to slice rather than spoon it out at breakfast.

Liz and I had long since given up trying to feel dry, resigning ourselves to a clammy existence. Even if we looked and felt like two ice-lollies someone had left out of the freezer, we could not let our discomfort hamper our study. In fact, the rains heralded a particularly important stage of the research, for Liz wanted to see if the added water would cause the giant otter families at Lama to alter their behaviour in any way. Would they relinquish their core areas for the five-month rainy season and make off into the flooded forests as Nicole Duplaix's groups in Surinam had done? Or would they perhaps change their habits in some other way?

As it turned out, nothing happened. All four family groups carried on just as they had done in the dry season. They visited and marked their core areas every two or three weeks, while Shadow, Spotted Dick and Co.'s courting transient continued to visit and be tolerated. Looking around her, Liz was not particularly surprised that the families had stuck to their old routine. The creek forests never once flooded, so there was no need for Spotted Dick's group or the Maduni group to desert their core sites. Creek water level depended far more on the tides than on the rains, some of the highest levels occurring in the dry season during neap tides. But even then, none of the marking sites in Dark Lane or Maduni were ever affected. Sites on Russel Lake remained dry as well, though not without a little human help. I had actually expected many of the sites belonging to the Streaky and Anadale families to flood, but although the water level rose steadily in the lake, the sluice gates were always opened long before the critical level was reached.

The fish in Nicole Duplaix's study area in Surinam had spread out into the flooded forests during the rains. But at Lama they were forced to stay put; the boundaries of their watery world expanded only slightly in the swamp and creek, preventing the water in these two habitats from mixing. Only in the savannah swamps were new pools created. Here, the water crept outwards from the deeper dry season ponds to cover every nook and cranny of grassland. But this was not a major transformation and, all in all, there was little change in the location of fish pantries where the giant otters fished at Lama. This probably explained why our otter families had not bothered to alter their timetable in any way.

There was no need, either, to seek unflooded marking sites; the vast majority remained high and dry. Dry in giant otter terms, that is. To a non-aquatic zoologist, the sites were pools of liquid mud. It was not so much the inundation from below as the torrents from above that made spraint-collecting a messy oper-ation for Liz. She tried to recapture her mud-loving days of infancy in an effort to enjoy rather than endure her working conditions, but the psychological gimmick did not work. After a hard day's slog, there was very little of her that was not mud.

The work continued well into May, week after week of dripping skies and drenched earth. Liz stepped back into the boat in Bee Canal one muggy afternoon after checking her fifty-ninth site for the month, and exploded. 'Lord have mercy! What the hell's a nice girl like me doing in a place like this collecting shit all day? We all have our burning ambitions,' she continued in mock despair, 'but this is ridiculous!'

Time for a dose of husbandly support, I thought, and summoned up my most reassuring sympathise-cum-revitalise-the-victim voice. Liz listened to my little speech with a straight face and replied in a humble voice, 'Oh, Great Mentor of Weak-willed Scientists, Great Stiffener of Wobbly Upper Lips, I hear your sage council and am filled afresh with the flaming fire of motivation. Even now I feel the urge to skip lunch and see what lies on Site 68!'

Trying not to laugh, I pretended to pick up an imaginary phone and performed the motions of dialling. 'Hello? Hello? Is this the Expedition Breakdown Service? I've got an emergency here, she's in a bad way. Yes, just suddenly, while seeking after the Truth. Yes, yes; uh-huh. . . . In the middle of a lake in South America. But what? You don't work on a Sunday? Oh . . .'

'Lunch,' grinned Liz. 'That's all this patient needs. Half a cheeze-bake and a big fat slice of pineapple!' The Great Mentor's stomach grumbled in agreement.

Liz joked, but she knew the value of continuing the sampling come thunder or lightning, high water or anything else. Already she was beginning to find that our giant otter friends ate and behaved very little differently from in the dry season. Sightings were just as frequent but usually shorter. The white walls of rain made it difficult to see more than twenty or thirty yards ahead so distant viewing was rare. Even when the deluge let up, the clouds shut out the sun and killed the tell-tale streaks of silver. But gradually, over a period of weeks, Liz managed to collect all the rainy season data she wanted. There was not much point in spinning out the season at Lama any longer and every reason for getting into Georgetown. Apart from the early downpour sequences I had taken, the rains had literally made heavy weather of the filming. The light simply was not there even if one could have protected the cameras from the worst of such foul weather. We decided that now was a good time to submerge completely and follow the giant otter's behaviour underwater. We would go to Georgetown Zoo and work with Squeaky, the captive otter, in her newly built, glass-fronted cage.

13 Squeaky

(Keith)

Georgetown looked completely different under a rain-sodden sky. With its pooled roads, its hundreds of umbrellas and glum faces, it bore more than a passing resemblance to London and the sights and sounds we had come so far to escape. As soon as we had unloaded our belongings we made straight for the zoo, accompanied by Nibs, who was now so grown that his beer-box sleeper was rapidly becoming too small for him. We had had no word from the zoo for more than four weeks and we were anxious to discover if Squeaky's enclosure had been completed before the onset of the rains. Prior experience warned us to expect the worst: the Guyanese contractor had already been working nine months on the cage, having promised its completion in three weeks! If the job was not finished now, it would have to wait until the end of the rainy season and that would ruin our schedule.

We left Nibs locked securely in the Superintendent's office with a pile of freshly caught patwa, and made our way to the back of the zoo. To our great delight the cage was standing four-square and solid, with the contractor putting the finishing touches to the wire bars. It looked perfect – twenty-two feet long and eighteen feet wide, with an eleven-foot-wide pool running the length of the cage. Behind the water was an area of dry land, seven feet broad and randomly planted with the bushes and plants we had brought from Lama on our last visit to the capital to give the authentic feel of creek habitat. The rain had at least been good for the vegetation: it was sprouting vigorously, making the cage indistinguishable from Dark Lane or any overgrown areas on Lama Creek.

The most important aspect of the cage, the pond, looked good too. Five feet deep at the front of the cage, it gradually sloped to a shallow area close to the land and gave ample scope for swimming and diving. In the front wall of the pond was a three by two feet sheet of toughened glass to allow for observation and filming of the otter underwater. We had insisted on toughened glass; a colleague at the BBC had narrowly escaped death when the water pressure in a similar pond had fractured his window of ordinary

141

glass and sent razor-sharp silica spears thudding into the wall behind him. One had hit him in the shoulder, causing a nasty flesh wound. We wanted to avoid any such repetition, especially as the glass would have to withstand a sixty-pound animal crashing against it once we had transferred Squeaky, the captive giant otter, to her new home.

Ten minutes after her transfer, Squeaky was swimming enthusiastically in the pond, chasing the live fish that Liz had thoughtfully added to the water by way of a house-warming present. While it was not possible to watch the giant otter's more typical shallow-water fishing behaviour in the cage, the glass sheeting allowed Liz to collect quite a lot of new data on open-water hunting. When the water was clear, Squeaky spent a lot of time floating on the surface or moving slowly about the pond using a lazy dog-paddle. She kept her head down during these periods and we could see that her eyes remained open as she scanned the water for fish. They bulged peculiarly from their sockets, giving Squeaky a slight look of Groucho Marx, but as soon as she surfaced, they flattened back to their slightly short-sighted contours. This outward curving of the eyeballs is an adaptation for seeing better underwater but not all semi-aquatic animals use it.

When visibility in the pond was obscured by mud and other debris, Squeaky never used the floating manoeuvre but simply dived immediately. She seldom returned to the surface without a catch, and we assumed that she sought her prey on these occasions using her well-developed facial whiskers as remote sensing devices. European and North American otter species have been shown to do this and they are still surprisingly efficient predators, even when fishing on a moonless night. In clear water Squeaky usually pursued her prey in a very direct manner: she simply made straight for a school of fish. This would scatter immediately on her approach, each fish making off in different directions in an ichthyological version of *sauve qui peut*. Squeaky seemed to pick her prey at the very last moment, typically going for the closest fish and pursuing it vigorously. The chase did not last long. The fish seemed to tire quickly (as they had with Nibs in the shower), although some did try to escape by twisting and dodging round the submerged tree roots we had introduced into the pond. We never saw Squeaky miss a fish; she normally took the quarry in her mouth with a swift bite. Twice she used her forepaws as an aid to catch a particularly agile fish. On both occasions her prey had just managed to evade the otter's jaws and would otherwise have escaped. Squeaky rapidly brought her forepaws into play, palming the fish into her mouth with all the skill of a legerdemain artist. On several occasions we saw that Squeaky used the same ploy that Nibs had used on his floating toys, swooping up on a fish from behind and below, like a fighter pilot in a dog-fight. Squeaky's prey seldom tried to make off when attacked in this way, probably because such an angle of approach is a blind spot for most fish.

Once caught, the fish were invariably bitten and carried to the surface. Here, with Squeaky resting half-submerged on a handy log, the meal was eaten. We had seen such behaviour often in the wild, especially in deeper water, and considered it typical of the giant otter. More rarely, Squeaky floated on her back to enjoy her food, growling contentedly all the while. At such times she was very sensitive to the proximity of people and we were twice charged by her for straying too close while she fed. This aggressiveness explained our observations that except for the 'wah-wah' calls of the youngsters, any giant otter with a fish was left very much to itself. On most occasions the captured fish was bitten heavily in the head and then eaten, always head first. To anyone who has tried to hold a slippery fish this makes perfect sense. The centres that control wriggling are all concentrated in the fish's brain. A series of bites to the head will almost certainly destroy these centres, so preventing the wriggles that might spoil an otherwise enjoyable meal.

Liz watched all these gory encounters with great enthusiasm. Not only was there behavioural data to be had from Squeaky's feeding activity, but Liz could use the same behaviour to learn more about the giant otter's diet. As she had done with Nibs at Lama, Liz had purposely fed Squeaky with live fish from several known species. Once the otter had sprainted, Liz could then take samples in the same way as she had for wild giant otter. Knowing the identity of all the fish Squeaky consumed, she could analyse the sample to see whether each species Squeaky ate really did show up. If it did, Liz could be sure that her conclusions about wild giant otter prey preferences, seasonal variations and so on were indeed correct.

To her obvious delight, the technique worked. Each fish eaten showed up in the samples that Liz religiously scooped up each day from Squeaky's latrine. But for me, life was pretty dull. I was dogged by the same problem that had prevented photography on Lama: despite the fact that the weather had brightened considerably in Georgetown, there was not enough light to film underwater. It was a great disappointment. Instead of exciting scenes of fish pursuit and capture, I had to content myself with filming

BCUs – big close-ups – of Squeaky's head when she came ashore to rest. To do this I had to cut the wire in front of the camera lens; it made quite a big hole, but not one large enough for Squeaky to use as an escape hatch. The problem about this set-up was that there are only a certain number of ways one can film an otter's head. After half an afternoon's work I had filmed Squeaky's cranium head-on, in profile, from above, when she yawned, as she dropped off to sleep, when she groomed, scratched, even when she defecated! There was nothing left to do. And until the rainy season ended, it seemed that that was the way it would stay.

Then, about a week after we had set up our position next to the otter cage, I looked up to see a tall negro, a Rastafarian by the length of his dreadlocked hair, staggering uncertainly towards the back of the zoo grounds. This was not unusual, since we often saw people suffering from a surfeit of Guyanese rum in the zoo. What was unusual was that the man was stark naked! He was about a hundred yards off when he suddenly stopped, next to the lion cages. He turned with theatrical care and stared at Simba, the large male lion, an enormous grin on his dark features. Then all at once, he had leapt the barrier and thrust his arm into the lion's den. I started forward but it was all over before I had moved even twenty feet. Simba seized the man's arm between his enormous forepaws and dragged it to his mouth. I saw the huge jaws crushing down on the forearm but the man's face retained its idiot grin, even when Simba released his bite and I could see that the man had been badly mauled. The Rasta staggered back, dragged himself over the barrier and weaved drunkenly towards the otter cage. By this time I had reached him, and I tried to guide him gently back towards the zoo entrance, where I knew there was both a first-aid kit and a telephone. But the Rasta pushed me roughly aside and before I could stop him had jumped into the canal which surrounds the zoo to prevent unauthorised access. He swam strongly, seemingly oblivious of his wound, and emerged onto the pathway with his arm dripping water and blood. I had no idea what to do. Liz was away shopping and I did not want to leave the camera equipment unattended. But equally, I could not let an injured man walk off without trying to help. At that moment one of the keepers appeared from behind the ocelot cages. Asking him to keep an eye on the cameras, I ran down to the zoo offices, gasping out my story and suggesting that we take the zoo van to see if we could catch the man.

To my astonishment, no-one seemed bothered. The man was 'only a stupid Rasta-man', I was told. 'He high on ganja, he no wort' boddering 'bout.'

It seemed that the zoo men had seen it all before. Rastafarians believe that ganja – better known in the West as marijuana – is a holy herb which helps them to find God. They also hold it as an article of faith that Haile Selassie, the late emperor of Ethiopia, is their spiritual father. One of the titles held by Haile Selassie was 'The Lion of Judah' and it was this association that had brought several Rastafarians to Simba's cage, though never with so serious a consequence as I had witnessed. So that was that. Without the zoo van there was little chance of finding the injured man. He disappeared and, although I scanned the local newspaper daily, no word of his fate appeared on its pages.

Such a morbid preoccupation only revealed how bored I had become. Any distraction was welcome. So when a gang of school-kids began shouting out questions about the camera during a particularly tedious afternoon, I jumped at the chance to go over and explain our work to them. I was just beginning to enjoy myself when the whole class gave a simultaneous roar. I spun round to see Squeaky worming her way through the gap I had cut in the wire for the camera, splaying the bars wider by simply pushing through them. At the same time I took in the fact that our precious Arriflex was teetering on its tripod as Squeaky forced her way ever closer to freedom. To a score of cries of 'de big wata dog loose!', the children scattered in all directions. I tore back towards the cage and was just in time to grab the camera before it crashed to the ground. At the same instant that I rescued the Arriflex, Squeaky finally managed to ease her hips through the wire. She slipped gracefully to the grass at my feet, panting through her open mouth. I had a moment of sheer terror. Memories of the Brazilian bitten to death by giant otters returned with horrible clarity. And now a six-foot giant otter was not a yard from my feet! I backed off slowly and carefully, holding the tripod in front of me as a shield. But Squeaky was not particularly interested in being vicious. This was the first time she had been out of a cage since she was brought to the zoo as a young cub some eight years before. Away from the confined world of the cage she was disoriented and very nervous. I began to feel that we could recapture Squeaky easily; all I had to do was to talk to her gently, to keep her calm until the keepers arrived with their nets. I squatted down, still with the tripod as a screen, and tried to smother Squeaky's disquiet with a blanket of soothing gibberish.

It just seemed to be working when the schoolchildren returned. They were in as high a state of excitement as Squeaky was. Anyone under ten seems to delight in being chased and they exult in the feeling of terror-panic that accompanies pursuit. These children were no exception, and now they had a real live ogre, a water dog, right before their eyes. It was too much, they simply could not resist baiting the animal, running in close and then tearing off with delighted screams of fear. These activities had the desired effect on Squeaky. She became progressively more excited and finally took off in pursuit of the yelling mob. They scattered in panic as the otter bounded towards them, barking in a mixture of fear, anger and excitement, her red mouth showing a trickle of saliva at the edges. Almost as soon as the children fled, Squeaky stopped. But the mob regrouped and, despite my calls for them to stop, began the baiting process again. Squeaky charged once more and this time it seemed she would not stop until she had one of these annoying creatures between her teeth. I snatched up a stick and caught up with her, trying to head her off from the children.

On the edge of my vision I could see a family of five standing pressed against the wall of the nearmost cage. I tried to move Squeaky away from them with the stick but she replied to my constant poking by biting it, snapping the stick in half. It was a nasty moment, and I was glad when I heard the sound of the keepers approaching. Two or three clouts round the ears from the head keeper, George Lee, soon dispersed the schoolkids and, faced with this new influx of larger humans, Squeaky once again came to a halt, looking round her dubiously. The keepers rushed her en masse, throwing nets. One covered her and George Lee pounced, trying to bundle the otter even more securely into the net. Unfortunately, Squeaky was less encumbered by the net than she appeared. She spun round and there was an audible crunch as she crushed George's hand between her jaws. Then she was struggling to escape, already half out of the net. Squeaky would undoubtedly have broken loose, producing more problems and suffering more stress herself, had it not been for the father of the trapped family. With commendable initiative, he seized a net that had been dropped in the mêlée and threw it skilfully right over Squeaky's head and forepaws. With her most important defences hors de combat, Squeaky was easily overwhelmed and carried, protesting, back to her cage.

After closing Squeaky's escape hatch with more wire, I went over to thank the amateur net-thrower for his help. He was a Chinese, of middle height with dark hair and sparkling blue eyes. He introduced himself as Neville Chin, a botanist working with the Guyanese Rice Board. Neville had spent some time in Britain at London University, and it was there that he had met Pauleen who was now his wife. They had three children, Jacky and Judy aged

146

16 and 11, and their 14-year-old son, Steven. I invited them to drinks at the zoo bar, and on the way there looked in on Nibs to make sure he was all right.

We never got to the bar. The moment they saw him, the whole family crowded into the room and began playing with and petting Nibs. The young otter was the star of the show, and he knew it. He rushed around nibbling toes, or rolled on his back and waited expectantly to be tickled. What surprised me was that the whole family knew that Nibs was an otter. When we had taken him for walks in Georgetown, Nibs had been variously identified as a dachshund, a seal, a caiman and even a snake! I mentioned this and Neville explained that they lived at Ogle Plantation, an agricultural station that backed Russel Lake and was linked to it by a series of canals. They often saw otters in these canals and so were quite familiar with them.

Judy looked up from stroking Nibs' long tail. 'What will you do with him when you go back to England?' she asked suddenly.

It was a question that had worried me not a little, and I took some time to answer. 'By that time we hope Nibs will be old enough to fend for himself, and then we intend to release him back to the wild.' I paused. 'The problem is, we don't know if he'll be able to look after himself when the time comes for us to leave. If he's not ready, we'll have to leave him at the zoo or take him back to England with us. I don't really like either option, but it's the best we can do.'

Pauleen didn't hesitate for a moment. 'You can leave him with us, if you want.'

'Oh, yes! Yes!' chimed in the children. 'Could we, Daddy, could we?'

I looked at Neville, afraid that this suggestion might have precipitated a family quarrel. It was one thing to admire a baby otter in an office, but no-one knew better than I did that to care for one on a twenty-four-hour basis was quite a different proposition. Neville was staring at the ceiling with the expression of a long-suffering martyr on his face. At first I thought he was really put out by the suggestion. Then I saw that his eyes were twinkling. 'Do you know,' he began, his eyes still locked skywards, 'that in the past two years we've played host to a caiman, a parrot with a broken wing, two tree boas and a three-legged opossum, and it was only two weeks ago that we finally managed to find a good home for a monstrous capybara that for the past six months had ruined my evenings with its nocturnal ramblings.'

A storm of protest greeted this statement. 'But he's only little.'

'He'll sleep all night, I promise!'

147

'Oh please, Daddy.'

'No!' Neville's word was final, his face stern. He held up his hand for silence. 'No, with experiences like that,' and his face broke into a grin, 'I'd say that a baby otter would be no trouble at all.'

The children all rushed over to hug their father and it was settled. If Nibs was not ready to take up an independent existence by the time we left for home, the Chins would become his foster parents until he was. The family left in high spirits, but I was doubtful that they would ever have the pleasure they anticipated. Nibs had already shown that he was capable of catching his own food, and he could swim in the open creek with astonishing confidence. Besides, there was still another five months before we had even to think about leaving Guyana. No, there was very little chance of Nibs spending time at Ogle Plantation.

14 Glimpsing the Pearly Gates *(Liz)*

Back at Lama, we spent the dry season doing routine work and more filming. Our only break came in December when we flew over to Barbados to spend Christmas with my parents, but the New Year saw us back in the familiar creeks and steaming swamps. Any belief I might have had that studying giant otters was going to be a continuous round of interesting experiences quickly evaporated in the humdrum routine that followed. But when excitement did come our way, it was all the more worth it for having endured the boring bits.

I experienced one of these high points in mid-February. Keith was staying with Coxy, the flagstaff watchman, at the time, filming one of his pet birds, so I was alone in the canoe as I awaited the arrival of the Streaky family in Lama Pond. The usual preparations had been made: gun-mike pointing directly ahead to record any vocalisations, binoculars on the ready and stop-watch primed to measure whatever behaviour I might be allowed to observe that morning. The family arrived around nine-thirty, just as the coastal breezes were beginning to blow across the lake. Two heads popped up across the open water – Mister and Missus – followed in quick succession by Dot and Scratch. A small movement caught my eye behind the bobbing bollards. It looked like a blob of moving oil but was much more alive than that.

'I don't believe it, it's a baby! It's a bleeding baby otter! And a female by the looks of it!' I muttered excitedly to myself.

It was swimming low in the water, just as Ringo and Spotted Dick Jr had done when they were small, but its head was even smaller than theirs had been and I reckoned it could not have been more than three months old. It kept well behind the four periscoping adults and every now and then vanished among the wildly flapping lily pads. Turning westwards, the family moved across my field of vision and made for one of their regular feeding-grounds that lay behind a dense web of weed and lily mats. It was hopeless to try following them through that lot, but the next day I was out again to see if I could catch another glimpse of Elka, as I called the little cub. The Streaky elders dived easily into the dark depths and at times they came so close that I could hear the soft expulsions of air as they surfaced. Elka struggled to emulate them but she never managed to stay down as long as they did. I wondered if this was because she was not yet very good at holding her breath or because she was finding it difficult to overcome her natural buoyancy. She tended to keep close to the lilies with the same show of reserve that had characterised Spotted Dick Jr.

I was startled when suddenly I heard Missus sneeze-grunt and look towards the lilies. Rarely did the Streaky group talk to one another while we were around, but here was Missus breaking her reserve and proving she had a larynx after all. Another two sneeze-grunts followed in quick succession – Mister was giving voice as well. Something was up but I could not fathom what it was. I saw the two of them swim into the lily patch, their heads twisting from side to side as if looking for something. Dot and Scratch did not seem too concerned by their parents' discomfiture and continued to hunt in the open water. It was the first time I had seen so great a distance separate parents and cubs. Two minutes later, I heard and then saw the reason for Mister and Missus' concern. Elka appeared from deep within the mass of lilies and floating weed, sneeze-grunting in short high notes for all she was worth. Her parents answered her and met their lost baby halfway across the floating leaves. They touched noses and bobbed heads in a touching ritual of reunion. Elka had returned to the fold.

The next cycle in March saw the whole family arrive in their lakeland core. Again they stayed for only two days, hunting for a short while each morning in Lama Pond. Nothing had changed; the family was still very much a unit. March passed quickly and my birthday came and went. I celebrated it with lentils and rice pudding over a damp sizzle of a camp fire on an island in Russel Lake that moved during the night. Not everyone's idea of high living, but it was memorable nonetheless and a helpful distraction against the slow plod of the Streaky saga. Ten days later, Keith and I were back at our usual watch-point in Lama Pond, waiting for the family to arrive. Ten-thirty passed and by eleven o'clock we were still waiting.

'They're either late or they're not coming this time,' yawned Keith, leaning tiredly against the tripod. 'What say you we move off in quarter of an hour?'

I nodded, and scanned the grey expanse in front of me for the hundredth time that morning. There was not a whisker to be seen and it looked as if we were in for a downpour. A band of dark cloud was coming our way and I watched as the first drops peppered the water, turning its surface from a smooth gloss to a dull matt. The wind gusted across the lake and caused a multitude of lilies to flap their leaves in noisy applause. We opened up our umbrellas and continued to look across Lama Pond for signs of life. A movement in the sedge to our right suddenly distracted me. Swinging round, I saw an otter's head not twenty feet away on our side of the pond. It was partly hidden by a narrow fringe of grass growing out of a dense clot of decaying sedge, but we caught glimpses of pink chin and dripping whiskers. There were loud crunches as the jaws crushed a broad silver fish, perhaps a catacari. The otter was slouched over the floating ramp, half in and half out of the water. Before the meal was through, another head appeared in the open water and periscoped. It was Dot, looking a little surprised to see us so close. True to form, she did not 'hah!' or snort, just checked us out silently and then veered off towards her partner who by now had slithered off the ramp, licking his lips. Dot looked at him and jerked her head back at us as if to say, 'It's them again!' But her companion wanted to have a look for himself and came towards us in a slow dog-paddle, periscoping within a boat's length of us.

'It's Scratch! They're alone! The family's split!' I burst out.

'Uh-huh, but would you mind moving out the way? Your head's blocking my view.'

I looked over my shoulder and caught my reflection in Keith's zoom lens. 'Terribly sorry, sir. There – the vista's all yours.' I squatted down in the bottom of the boat and continued my frenzied scribbling, umbrella in one hand, notebook balanced on knee. 'No sign of Elka and her parents,' I wrote. 'Dot and Scratch not following the usual travel pattern, heading instead in the opposite direction through the sedge on the north of Lama Pond. Out of sight now, but making spasmodic sneeze-grunts and very intent on fishing.'

We decided to try and follow the sounds but no sooner had we undone the mooring than a blast of coastal wind swept us across the churning water. My efforts to control the boat were useless and before we knew it, we had skimmed over to the other side of Lama Pond and now lay floundering among the weeds and lilies.

Keith downed camera and stepped to the back of the boat. Leaning over the edge, he pushed with his paddle against a thick ball of matted weed while I went to the front to paddle and steer. We grunted with the effort of pulling and pushing against the wind and waves but it got us nowhere.

'Keep us into the wind, for Chrissake!' Keith yelled.

'What d'you think I'm bloody well trying to do!' I spat back.

The boat twisted and turned but the wind kept flinging it back into the jungle of water plants. I turned round to tell my cursing crony that it would be better to wait it out until the elements calmed down. In that instant, a sudden squall slapped the canoe sideways and upset Keith's centre of gravity. His body stood poised for a hopeful moment, coming within an ace of recovery, but the wind tipped the balance in a very literal fashion and sent him plummeting into the depths. I remember thinking, 'Thank God the camera isn't with him this time!' We had already lost the Olympus and our precious 300mm telephoto lens that way and our insurance premium was rising with each dunking.

I came to life and got ready to haul Lama's water baby back on board. I waited ten seconds, twenty seconds, half a minute and still no Keith. What had seemed a rather funny incident suddenly turned serious. I was frightened now. Lying flat on my stomach, I hooked my feet under the back seat and leaned over the side of the boat. From above, everything was invisible. I dunked my head and shoulders into its grey anger and looked for Keith, heart pounding in fear. I saw his blurred form about four feet away. He was struggling to free himself from the long stems of the water lilies that had caught his legs. Without looking back, I fumbled for the sheath that was strapped round my thigh and pulled out the knife I used for scraping up otter spraint. Dragging on the same lily stems with my left hand, I brought me and the boat within reach of Keith and pushed the hilt into his hands. Frantic now, his lungs aching for air, Keith slashed out at the ghastly green ropes, cutting his thigh in the process. He broke surface with a long, life-giving gasp, and gulped at the air again and again, oblivious of everything else. But I was terrified lest the blood pouring out of his thigh would attract pirhanas and start off something more horrific. I dragged him bodily into the boat and took a quick look at the wound through the slice in the trouser-leg. It was deep – right through the fat and into the muscle – but it was a clean cut and would heal if we kept the two halves bound together.

'Dear diary,' panted Keith, his face slack with exhaustion, 'caught a glimpse of the Pearly Gates today. They looked lovely but the journey there was lousy.'

'Talking of journeys,' I interrupted, 'it looks like it's clearing up. Like the good fairy I am, I'll paddle you back free of charge. You just lie there and talk to yourself in luxury.'

Keith complied, but he refused to let his gash hinder our work in any way and the next day saw us out on Lama Pond again. But

Dot and Scratch did not come back. We had no idea when they would return to Lama Pond and we could not afford to give them any more of our time as this was something we were running short of. The study was coming to an end and there were still plenty of things to see to. Carabice Creek, for one, still had to be surveyed.

Carabice was a long, winding waterway that formed part of a dense network of creeks on the southern boundary of Russel Lake. Iburu and Ram Raj had both gone on about the vast number of big water dogs that lived there and on their extra-aggressiveness. 'Dey bold, bold, dem wata dogs. Dey rush our boat many times and we had to mek noise wid our paddles to chase dem away.' We could not tell how much of this was truth and how much the desire to impress, but we decided it was worth checking out all the same. To get to this tribe of furred warriors we were told we would have to paddle for about seven miles across the lake. Then the creek itself wound for another ten miles or so before joining up with the clearwater creeks in the Timehri sandhills.

'Engine no good in dey,' both men warned. 'Takoubas everywhere. A big one wrench off our engine two week ago and brek de pr'eller blade. You got fuh be very careful.'

We took their advice and paddled. Engines and engine parts were precious commodities to Guyana because neither was imported. There was a big demand for them and few people would hire an engine out for fear it might be returned damaged. We decided to bring ours along just in case we found we could use it in the deeper parts.

The mist was still languishing over the water when we started out. We brought Nibs along too because we wanted to try and harden him slowly into coping with strange places before returning him to the wild. We fixed an extra-long lead on him and he swam obediently in the water beside us, never straying very far from the boat. But he slowed us down at first by insisting on playing dodgems with the paddles as they dipped into the water. It was too early in the day to remonstrate him. Better, we thought, to conserve our energy for the day ahead and let him tire of the game of his own accord. He did this remarkably quickly when he realised that the paddles did not tickle or taunt. Weeds were far more responsive and they also had the odd chaseable fish tucked away among their fronds. The only time he pricked his head up was when we passed by a noisy aggregation of parrots clinging to the oily fruit of an Ite palm. They clustered round the feast in layers of flapping feathers, each bird jousting to keep a foot on

153

the communal table. Nibs listened to the squawks and squabbles with frightened eyes and dived back into the safe walls of water.

The route to Carabice was exhausting. There were endless twists and turns to negotiate which made my steering muscles burn with fatigue. The waterway narrowed to a canoe's width at some points and at others it was completely blocked by fallen Ites. Keith hacked at these to clear the way. It was out of the question to try and heave the canoe over them; not only was it too heavy, the channel was far too deep for wading in. We could just make out the trees of the Carabice in the distance, standing out green and cool from among the vast expanse of brown sedge and lonely palms. According to Iburu, we had to pass through two small lakes before we reached the creek entrance. The sun was almost overhead by the time we came upon the first one. It seemed vast compared to the thread of water we had just navigated, big enough for the wind to ruffle its surface into small waves. I looked at it in a state of detachment, having long since disengaged my mind from my body. I had done this many times during the trip and on other expeditions. Past experience had taught me the value of switching off hard exertion or physical discomfort, of thinking about something else other than hunger, thirst, muscle ache and burning skin. I was well and truly hypnotised by the time we broke out onto the lake so I did not see the head moving across the water right away. It was just a spot in the corner of my eye. But it grew bigger and eventually forced itself upon my consciousness. The otter made no sound but I knew that it was not an apparition because Keith was looking back at me and pointing at it. Nibs suddenly added his support to its solidness by wheep-wheeping loudly in fear. I obligingly yanked him into the boat by the scruff of his neck, pleased to see that his instinct for self-preservation was in good working order.

The giant otter had disappeared silently with Niblet's first wheep. 'Nothing aggressive about that one,' I said. 'As shy as any other transients we've come across.'

'We haven't reached Carabice proper yet, though,' Keith pointed out. 'May be all sorts of crotchety otters there ready to do us in.'

'Well if there are, I don't think I've any strength left to fight 'em so you'll have to protect me, Tarzan.'

'Sorry, Jane, but Tarzan no longer protect females. New Union rules.'

'I'll scream pathetically and faint.'

'No good, I'm afraid. Masculine Liberation Movement – I'm allowed to faint too. Try bribing me.'

154

'Okay, I'll pay you overtime, and make all our coffees to-morrow.'

'Done.'

We used this sort of silly banter from time to time to augment the mind-detachment ploy. It provided entertainment on arduous journeys and helped keep our sense of humour in good repair. But I certainly hoped Ram's description of the otters' mean dispositions was a joke.

There were no other aquatic sights of interest until we entered Carabice itself. The water was much shallower here and as we paddled round the first of many bends, we could actually see the brown mud bottom and the grotesque form of each takouba. Ram hadn't been joking about those, at any rate. We took half a dozen bends in as many minutes and were about to dig our paddles in for the next one when there was an almighty chorus of snorts and three giant otters catapulted off the bank, plunging into the water on our starboard side. The creek foamed and sent ripples in all directions. Taken completely off guard, we froze for one second too long and the canoe rammed to a jolting stop. It was a massive takouba and it had us securely by the middle. There was nothing we could do just then so we postponed the problem and looked up and down the creek for signs of the otters. Not a bubble. But our field of view was restricted to a small arc of creek so it was possible they were round the next corner. We would have to get ourselves off the takouba first before we could follow them.

We had been grounded before by takoubas but never so tenaciously. Keith went to the front of the boat while I remained at the back and together we tried to swing it off its stubborn pedestal. It did not work and each gyration produced a sickening sound of wood under pressure. Too many swings like that and we would hole our precious vehicle. So we then both stood at the back in an effort to raise the front end, poling the mud with our paddles. That failed, too. Nibs did not make the job any easier by treating the takouba and its branches as slalom poles. He weaved in and out of them with the fluid ease of a fish, pleasure written all over his little face.

'Niblet, for Pete's sake, get away from there!' I shouted. 'You'll get crushed. Here, tch-tch-tch-tch. Eat that.'

I flung the patwa towards the opposite bank to distract him and Nibs, hearing the food sound, leapt after it. We decided to get wet and try to heave the boat off the fulcrum, but as it turned out we did not have to exert ourselves in the least. As soon as we got out onto the soft mud, our ship moved off of its own accord.

'Thank heav . . .!' Keith never finished his praises. They were interrupted by a fearsome roar of caterwauls and squeals that came from the moca-moca behind us. I jumped visibly and twisted round in the water. Through the commotion, we could make out the shrill squeaks of an animal in distress. It was Niblet. He darted through the water towards us, followed a few seconds later by three dangerously roused giant otters. Their mouths were agape, red and angry, their jaws opening and closing with the fury of their screams. I felt an exquisite fear pierce my stomach. Keith and I leapt into the boat as one, Nibs following hot on our heels. He burst from the water like a guided missile and cleared the boat with inches to spare. Fuelled by the adrenalin that coursed through our bodies, we grabbed whatever weapons we could. Keith raised the cutlass for the blow he might have to strike and I grabbed the heaviest paddle. The giants kept coming with awful certainty. It was then I remembered what I might do to halt them. Clutching my throat, I caterwauled with all the power I could muster. Keith looked at me in horror, thinking I had cracked, but he quickly realised what I was trying to do and joined in. We belted away in fearful desperation, giving as good as we were getting. Whether or not those huge heads would have stopped anyway, I will never know, but they did. They slowed right down and started zig-zagging around the boat, still blasting us but at a lower octave. Then after what seemed like an eternity, the head defender snorted twice and made off in the direction he had come. His mate and offspring followed him and the trio disappeared round the bend, sneeze-grunting and humming among themselves.

I dropped my paddle and sank down heavily, not quite believing it had happened. My whole body trembled with reaction.

'Oh my God,' Keith gasped, 'I thought we were done for there.'

We found Niblet crouched under my seat at the back of the boat. He looked as cowed by the experience as we felt and refused to come out when I stooped down beside him. Drained of his usual ebullience, he allowed me to untangle his lead without so much as a playful nuzzle.

'Sorry, old son,' I squeaked hoarsely, as I tied it short through a hole in the seat. 'We can't have you going on the rampage again.'

But Nibs had temporarily lost his desire to explore. He remained within his dark cave for the rest of the afternoon, refusing to come out even when Keith gave the magic food call. We had to bring him his lunch which he ate in close, reassuring contact with my feet. Looking down at him so innocent and vulnerable, I wondered what he had done to spark off the tempers of the giant otter family. Probably nothing. His only fault, or rather mine for throwing the patwa, was being in the wrong place at the wrong time. The fish – and Nibs – must have landed too close to the giant otters' hiding-spot and they had become alarmed. For all their threatening behaviour, there was nothing to suggest they were any more aggressive than other giant otter families, though perhaps the confining narrowness of Carabice made escape unusually difficult and encouraged conflict with humans. We would be the last to deny the experience had been frightening – Ram and Iburu certainly had my sympathy there – but the attack had probably been nothing more than a keep-off warning, a well orchestrated bluff-charge.

After all the excitement, the rest of the journey sank into anti-climax. We drifted through some fairy-tale scenery as soothing as the attack had been upsetting. The trees above were shedding their flowers on the water – masses of delicate yellow petals dancing onto the black glass of the creek water. Shafts of sunlight leaked through the branches invitingly and there was a sudden splash of silver as a kingfisher speared the golden carpet. It was a scene straight out of *Swan Lake* – softly beautiful and yet vibrant. The water itself was humming with life. Insects abounded on it and above it but, thankfully, none of them belonged to the blood-sucking squadrons of mosquitoes. During one of our short rests, I singled out a whirligig beetle from its fellows among the lilies and tried to follow its dizzy jig without losing track of it, but the task was an impossible one. I was giddy in a few seconds. This was probably how one fellow had managed to lure a small spider down on the water. The spider fell into the centre of the whirligig's mad orbit and lay there, dazed and unmoving. I was not sure whether the whirligig had thought the poor thing was worth eating or whether it just wanted to have some fun, but it suddenly grabbed the spider and took it on a whirlwind tour of a nearby lily leaf. The spider looked as if it was going to be centrifuged to death, but it stayed the course and eventually the whirligig ran out of energy and released it. I left the spider alive and well, if a little dizzy, having cadged a lift on a passing twig.

Just then, I wished we could have thumbed a ride in the same way. We had travelled seven miles into Carabice but, save for our hair-raising experience of three hours before, we did not find much in the way of hard data there, just two marking sites. One was large and extremely fresh, the other a tiny box of a bed-sit, with a small loo, an equally modest 'bed' and an unusually long access pathway. We turned back before it got too dark.

By the time the canoe slid out of Carabice and into the roofless

157

expanse of Russel Lake, the sun had lost its harsh, laser-like quality. I took the opportunity to give my flea-bites a dose of the soft rays and stripped off all my clothes. Fleas were the KGB of the insect world at Lama, experts at shadowing and persecuting warm mammalian bodies twenty-four hours a day. It was ironic, I thought, for someone to have to travel halfway across the world in order to catch common or garden dog fleas, but it was so. Twice a day I turned my clothes inside out to try and kill them but they always eluded me, either hopping off without my noticing or else burying themselves among the seams. Relief from the bites came only with nakedness. But this did nothing for my self-esteem as I looked like an inoculation experiment gone wrong! The bites stood out like beacons over the length and breadth of my body, their colours ranging from a bright burgundy to the blue-green hue of a corpse. The scars lasted for weeks and one sure way of giving them a long life was to scratch. Unfortunately, I did this all too often, such was the intensity of the itch. I had already scratched one too many today while paddling up Carabice Creek.

But now, the cool lakeland breeze brushed my skin and soothed away the itches. As I relaxed, my thoughts began to drift.

'Liz!'

'Hmmmmm?'

'I think I hear voices in front – men's voices.'

'What? Voices? Oh, Lord, brake! Brake!' I hissed in panic, dropping my paddle like a hot cassava. Niblet opened one eye and looked up at me as if to say 'What's she screeching at now?' But all I was concerned about was getting my clothes back on as quickly as possible.

I had no sooner done this, give or take a few trouser buttons, when a kurial rounded the corner with two bare-chested Amerindians in it. Their faces, gaunt and immobile, matched the dried-fruit texture of their bodies. We raised our hands in greeting but they just looked at us through shadowed slits. There was not so much as a flicker of social response, not the slightest shift in bearing. For some reason, this silence impressed me rather than annoyed me. These two ancients struck me as bearers of hidden thoughts and esoteric secrets: the stuff, I felt, of which hermits and philosophers are made. I had always wanted to meet a hermit and the middle of a lake in South America seemed as appropriate a place as any.

'Nice piece you got there.' One of them had spoken! And what was more, they were grinning from ear to ear, exposing battlements of blackened teeth. I thought at first they were referring to our outboard engine which was leaning against the seat in front of me with the handle sticking up in the air. Fishermen and hunters often stopped us to admire it and to ask 'how much for to buy?' But the gazes were brazenly directed at me and I heard Keith reply laconically, 'Yeah, not bad.'

I was astounded, but I was not to be outdone. Scratching at a flea that had re-colonised its territory under my armpit, I asked, business-like, 'How much for to buy?'

The wrinkles on their cheeks folded back like curtains and they both chortled with laughter, slapping their paddles in appreciation. Philosophers indeed, I thought. More like latent sex-maniacs. Trust Guyana to deliver the unexpected in the middle of nowhere.

15 Kaieteur
(Keith)

The flight from Timehri Airport to Kaieteur Falls takes a little over half an hour. The small plane shudders its way down the tarmac then suddenly the bouncy, bumpy ride ceases and an incredible vista of forest and river opens up as the aircraft gains altitude. The pilot banks the plane to the south-west and the mighty Essequibo River comes into view, almost ten miles broad at its mouth, and still all of three miles wide at the town of Bartica some fifty miles inland. Here the Mazaruni River joins with the Essequibo after its long journey from the base of Mount Roraima. The Essequibo is navigable up to Bartica, but thereafter both it and the Mazaruni are nothing but a mass of small islands, with the river water threading a tortuous, shallow path between them. Travel by boat is often possible only during high tides when the moon's pull gives the river added depth. As ever when flying over Guyana's interior, I marvelled at the determination of the Dutch, French and Englishmen who opened up the country. Not for them an effortless glide over the rain forest; they had to canoe and cut their way across the face of this harsh land.

The area above Kaieteur Falls was to be our last survey area. The time taken in filming, caring for and studying Nibs, and the wealth of useful information to be extracted from the giant otters around Russel Lake, had put our schedule so far back that it was now impossible to visit the Rupununi savannahs in the time still available. In a way, I was not sorry to have missed the Rupununi. Everyone we spoke to from that region had told the same story: that the giant otter had been shot out to supply the voracious appetites of fur traders on the Brazilian side of the border. It would have been best to confirm these stories personally, but the sheer unvarying weight of the tales pointed to their truth. I began to think of the Rupununi as a very depressing place. Having spent so long at Lama, the idea of a Guyanese river without a noisy family of giant otter was a sight I would rather avoid.

Seated comfortably in the twin-engined Otter (a good omen, we hoped!) the time passed pleasantly enough. We were becoming old hands at bush flying now. On previous trips, every creek, ox-bow lake and mountain, every circling vulture, flight of ibis or flock of parrots had been a point of interest. But now we were becoming quite blasé about such sights. We even felt that we had seen as much as we wanted of waterfalls; after all, there were no less than fifty named falls and rapids on the Essequibo River alone, and the forty miles below Kamarang on the Mazaruni boasts no less than eight of these major impediments to water travel. Falls, we felt, had become quite old hat.

This certainty vanished on our first sight of Kaieteur. We had read the descriptions of the falls and seen photographs, but nothing we had read or seen prepared us for the spectacle that appeared on the edge of the right wing and slipped beneath us as the pilot turned the plane for landing. The thing was colossal. Ton upon ton of water crept unwillingly to this huge geological step, surging over its face in an endless stream and dashing itself on the rocks 741 feet below. A permanent rainbow, generated by the sun and the spray from the falling water, stood like an ethereal, multi-coloured bridge at one side of the waterfall. When we landed, the falls were even more impressive. Kaieteur simply radiated power. It dwarfed us puny humans, and I could understand how easy it must be to return to an animistic religion where every grove, rock and stream has its own spirit. In such a creek Kaieteur would surely rank high in the pantheon. Watching it, I had that same intuition that comes from staring at the stars on a summer's night in England: the feeling of Man's total insignificance in the life of the Universe, and the pointless pettiness with which we fill our short time on earth. We are Lords of the World, but in the vastness of the cosmic hierarchy we rate about as highly as a single atom in the pattern of life on earth. We have our place, but it is an insignificant one, and places like Kaieteur reveal that truth with far greater clarity than any words. One's initial, egotistical belief that this great waterfall was somehow created solely for one's own benefit dissolves into the realisation that Kaieteur has existed for ages, long before the first human ever set eyes upon it. And if, as seems increasingly likely, Man's petty quarrels obliterate our species from the face of the earth, Kaieteur will continue to send its water crashing down into the Potaro River, supremely indifferent, as always, to the fate of humankind.

We had consciously to draw ourselves away from this mighty example of Nature's power, away from thoughts of universal harmony, and back to such mundane questions as where was the first spraint site. I had brought no ciné equipment on this trip; it was solely in the nature of a reconnaissance, a brief survey to establish the presence or absence of the giant otter on the Upper Potaro River. If the area proved an exceptionally good spot to film in, we would have the cameras brought in on the next plane.

As at Lama and on the Mazaruni, we used our normal survey routine of selecting likely areas and canoeing our way along them, looking for sight or signs of the river wolf. We did not have as many otter sightings as on the Mazaruni, but one observation was particularly interesting. We were canoeing at the mouth of the Kwitaro River during the mid-afternoon when Liz stopped suddenly and pointed upstream. I looked, and there was our first sighting, a singleton, a solitary giant otter moving out from the bank into midstream, swimming away from us. No, I was wrong; there was a second otter. No, a third. A fourth appeared, and we could tell from their relative sizes that it was a family group, two parents with a pair of two-year-old cubs. I looked away for an instant and when I looked back there were another two tiny heads swimming close to the adults. The parents swam steadily on, looking back towards us suspiciously and occasionally barking an order to their charges, presumably a command to keep swimming as fast as possible. They seemed quite nervous of our approach, and we could never get closer than a hundred yards to the group. Nevertheless, we were very excited by this sighting. This was only the second time we had seen a sextet in Guyana, though similar-sized families seemed fairly common in the area studied by Nicole Duplaix in Surinam. The younger cubs looked about a year old, which meant that there must have been only a year's gap between the birth of the two pairs of cubs, as against the two- to three-year gap between successive births at Lama.

Why this should be we were not sure, but it probably had something to do with a difference in food supply and other factors in the two habitats. The better the habitat, the more frequent the birth of young. If that was true, then it meant that the Upper Potaro River was ideal otter country as far as food, breeding sites and, probably, interference were concerned. There were many camp sites, too, so that even if we had not seen as many giant otter directly, we could have inferred their presence from the sites alone. It all added up to a picture of a healthy giant otter population on the Upper Potaro.

It was at Kaieteur that we said an unexpected goodbye to Nibs. He had become overheated in the canoe and Liz had dropped him overboard attached by his collar to a long nylon lead which she tied to the canoe seat. Nibs loved this and would quite happily dog-paddle alongside the canoe, keeping pace with our paddling. He never seemed to tire and even after we had shipped paddles in exhaustion he would keep up his leisurely swimming pace as he circled our now stationary canoe. Every now and then, on spying a particularly attractive prey, the canoe would yaw wildly as Nibs sped off in pursuit, only to be pulled up short as he reached the end of his nylon tether. We had just experienced another of these boat rockings when I noticed that Nibs' head had appeared on the far side of the boat, and quite far away. Too far away! I grabbed the leash and hurriedly dragged it aboard. It came easily, but at the end there was only Nibs' collar. The old leather had finally given way under the strain of Nibs' excited underwater pursuit, and our little otter was now completely wild and free in this strange river. I was full of ambivalent emotions at this unforeseen turn of events: pleased because I had dreaded having consciously to decide when was a good time to release Nibs, but very anxious because of my gut-feeling that today was not the day for him to strike out on his own. He simply wasn't fully ready for an independent existence yet. But how could he resist the lure of the creek? He had recently shown himself more and more frustrated by his leash and now, free at last from its constraints, it did not seem likely that he would give up his new-found liberty.

Nibs had by this time swum to the far side of the creek and disappeared into the green curtain of overhanging branches. We paddled furiously for the spot, calling him continuously, but when we finally reached the trees Nibs was nowhere to be seen. Even the 'tch-tch' sound brought no response. I was now full of foreboding. If he did not return, I felt certain that he would never find enough to eat. In a very few days Nibs would be dead.

We sat at the spot for just over an hour, calling all the time, and occasionally paddling a few hundred yards up or down the creek in the hope of bumping into our little lost otter. But after two hours, we had finally to accept that Nibs had decided he was old enough to look after himself. I only hoped that he was right and that my natural, though unfounded, misgivings were simply parental worries. Perhaps he would be OK. We gave one last look round, then turned the canoe for our camp site.

At that moment, the vegetation about ten yards upstream began to shake slightly. We had often detected giant otter from this sign, but equally, the movement could come from a caiman, a large fish or even a lizard or snake. Could it be Nibs? We delayed our departure and watched the creekside bushes carefully. Every so often the plant stems would move, and the movement was coming closer and closer to our canoe. I looked behind and Liz opened her hands in an expressive 'Don't ask me!' gesture. I turned back and, as I did so, Nibs' head broke the surface about

163

five feet in front of our boat and he stared at us with an unblinking, and I thought unknowing, expression on his face. I was torn between a desire to call him over and my partial confidence in his ability to care for himself. Before I could make up my mind, Nibs solved the problem for me by diving. I scanned the creek on either side of the boat for signs of Nibs, but he was nowhere to be seen. He had gone.

Then I heard a 'hah!' and a gasp of surprise behind me. Turning I saw Liz's smiling face and just in front of her, on the gunwale, one very tired young otter climbing aboard. Nibs was pretty well all in. He had obviously been in the water a little too long, for his coat was just beginning to look waterlogged. There was a little gash on his lower lip, his nose was green instead of brown, and water-weed stems hung between the toes of his hind feet. Nibs looked around him with the tired but satisfied air of one who has just swum the Channel. It was obvious that he had never doubted that we would wait for him while he went exploring. He dropped into the bottom of the boat, dried himself briefly on a canvas bag and made a bee-line for the fish-box. Liz fed him all we had, four red piranha, and he consumed them all ravenously; the tail of one had scarcely disappeared down his gullet before he had started on the head of the next. For all his two hours in the water, it seemed that Nibs had been singularly unsuccessful as far as hunting was concerned. If he could not find his own fish, it would be totally irresponsible to set him free at this time. As he finished his fourth fish, I made ready to catch him round the neck with a noose I had fashioned from the end of his nylon leash. But I did not have to bother: instead of leaping briskly overboard again, Nibs crawled slowly under Liz's seat, curled himself comfortably and was instantly asleep. For the moment, the big wide world was just a little too much for our prodigal son!

16   The Private Life of the River Wolf   *(Liz)*

By now there were only three weeks of the trip left in which to make observations of the family groups in and around Russel Lake. Looking back, I was pleased with how the study had gone. I had discovered much about the giant otter that would help governments and wildlife organisations to conserve it. The animal's environmental needs, its social set-up, its territorial policies and the extent of its home range were all now less of a mystery. Without guidelines like these, any effort to protect a species would be as ineffectual as trying to build a house without a blueprint.

My estimates on the size of family cores and the areas of home-range overlap would have an important bearing on any plans to protect the giant otter. A total range of twenty miles of creek or eight square miles of lake sounds vast but, like all big carnivores at the top of their food chain, otters have to live in low densities. According to Nicole Duplaix, however, giant otters live at extraordinarily high densities in the primary forest habitat of Kaboeri Creek in Surinam. On average, each family group survived in a two-mile stretch of water throughout the eight-month dry season. This gives a density of two giant otters per mile of creek, an extremely high figure for most smaller otters and without doubt far greater than anything we experienced even in the Mazaruni, which appeared to be a high-quality otter habitat. Account must be taken of this wide splay of spacing patterns in deciding, for instance, the minimum size of nature reserves. If we abide by Nicole's high-density figures only, we might make the reserves too small. The number of giant otters inside will then be too low to maintain a viable population, and far from conserving the animal, we would have fragmented the population and hastened its extinction.

It is no use a carnivore having a large range at its disposal without exploiting the area effectively, and the regular pattern of family group movements at Lama bore this out. Every two or three weeks, the families returned to the same spot to fish and re-mark. Sam Erlinge, the Swedish naturalist, found the Euro-pean otter does the same thing in inland waterways but there the pattern changes in winter when ice restricts hunting to certain stretches of river or lake. For a large fish-eating carnivore like the otter, cyclical movements are a reflection of good husbandry, as to remain in one place would lead to overfishing and eventual starvation. A giant otter family of four needs something like thirty to forty pounds of fish a day, more if there is a lot of travelling to do. Were the family able to catch any sort of fish, it might be possible for it to remain in one area for a long time but, in reality, a giant otter's diet is limited to certain types of slow fish, and this consideration forces the carnivore to keep on the move.

Closely bonded as family members are, they do not actually co-operate in hunting. I saw no co-ordinated fishing strategy, nor was there any food sharing. But if giant otters do not hunt in an integrated fashion, what forced them to evolve into social beasts? The only answer I could think of was predation: perhaps giant otters defend themselves more effectively as a group than as individuals. The anaconda may well be the biggest threat. According to the fishermen and hunters around Lama, ana-condas have been known to attack and drown single giant otters, and there are stories of pet cubs being snatched from the banks of Amerindian villages by the same villain. On the other hand, none of these men ever saw an anaconda take a cub from a family group which rather suggests that there is safety in numbers. The remarkable mobbing incident we saw in the Mazaruni supports this theory.

Predation could not, however, explain the formation of giant otter super-groups. Of these well-documented aggregations we had caught not a glimpse throughout the entire study. Latter-day naturalists had excitedly described close encounters with groups of up to twenty individuals, and Nicole had also seen super-groups of up to sixteen animals. No-one knew if the animals were all related to one another or when and why they formed. The remarkable thing was that super-groups should form at all in a species that spent most of its time avoiding one another. Perhaps these aggre-gations occur at times of food abundance. It is well known that European otters mark less frequently when food is plentiful, with the result that mutual avoidance loses its edge and chance encounters are more likely.

With only three weeks to go before leaving Guyana, we had our one and only rendezvous with a super-group at Lama. We had arranged to spend the day in Russel Lake with Iburu to search for takouba worms, a delicacy much vaunted by the locals. These creatures are not actually worms but the inch-long larvae of a mega-sized black weavil that builds its nest in the trunks of rotten Ite palms. It took us a good hour and a half to reach the first group of palms that Iburu had chopped down months before to encourage the insects. Our guide hacked at the first one with his cutlass, exposing the bright orange fibre inside. A black beetle suddenly scurried out of one of many corridors it had made in the fibre and played dead.

'You soon going see de worms,' Iburu assured us. On reaching the middle of the trunk, he stopped chopping and began gently to pare off the orange fibre. 'There,' he said, straightening up and wiping his face on his sleeve, 'dem takouba worms. Got plenty, plenty meat on dem and,' he said, turning to me, 'dey very good for make babies.'

I looked at the pulsating blobs and shuddered. 'Let's hope Keith and I can made the grade without that lot,' I said.

'But mistress, you got fuh try dem. Dey good fuh you!' Iburu insisted, with a rather hurt expression. 'I going fuh make a fire and cook dem while dey fresh. Dey better roasted.' He plucked the bloated albino grubs from their cradles and popped them one by one into a tin moistened with damp soil. The thought of eating them at all, raw or cooked, made my stomach heave but I did not want to hurt Iburu's feelings by refusing to try them. I turned to Keith surreptitiously and made a 'vomit' face. He was looking just as repulsed and whispered, 'This is one aphrodisiac I could do without!'

Iburu soon had a fire going. He positioned three stones round it and placed his heavy metal tawa, a portable cast-iron hot-plate, on top of them. Grabbing a handful of worms from the tin, he threw them onto the greased surface with practised ease. The poor creatures jerked once and died, shrinking in a few seconds to crispy commas. I tried to pretend they were tandooried shrimps but my imagination kept letting me down. Iburu flicked half a dozen into a calabash gourd and passed it over to Keith and myself. I waited in cowardly suspense as Keith tried the first one.

'Hmmmm,' he said, munching with a stoically sober expression. 'Delicious.'

Iburu then looked at me, waiting for my verdict. I picked one up gingerly and dropped it fatalistically into my mouth. I felt the hard outer skin crack and realised I was through to the soft, semi-liquid interior. Fighting nausea, I stopped chewing and swallowed. Iburu was watching expectantly.

'Oh yes,' I managed to smile, 'very tasty.' I was only glad I did not have to specify what it tasted of!

Iburu's face broke into a huge grin. 'See, I tell you so, mistress. Not one body I know who don't like takouba worm.' With that, he helped himself to a huge handful from the tawa and chewed them in the manner of a tanned Dr Who eating jelly babies. With my host's attention thus diverted, I deftly disposed of my helping by a tried and tested method, one to which we had often resorted on our travels around the world. Exotic dishes like boiled pangolin, casseroled rat, bat and frog had all been consigned to our capacious pockets. It was by far the best thing to do as refusal in most cases would have been tantamount to insult. Sometimes, though, we would forget the cache of rejected meals until we discovered a hideous ensemble of limbs and dried flesh when searching in our pockets for a light meter or camera lens.

We spent all morning gouging out rotten Ites, and by the time we came within sight of the Maduni sluice gates, Iburu had collected no less than half a bucketful of the wriggling horrors. He looked very pleased with himself and cheerily volunteered to fill up the engine with petrol when suddenly it coughed and died out. There was a marvellous silence in which my ears picked up the sounds of breeze and birds, but I suddenly realised I had tuned into a faint 'hah!' as well. Squinting over Iburu's bent head, I saw two otter heads appear from among the lilies, the large leaves slipping down their noses like loosened yashmaks. They periscoped and I recognised them at once: it was the Maduni group – Tom and Jerry and their parents. It was not unusual to find them in these parts as this was the overlap area they shared with the Streaky group. But I became very excited when I saw another two heads pop up within a boat length of the Madunis, followed by three others further behind. The pair periscoped and revealed strange chest patterns, but the threesome remained low in the water, keeping their identities hidden. I was looking at my very first super-group.

They shifted positions, disappearing and reappearing restlessly but always careful to keep the distance between us. It was obvious the association was loose as each family swam as a distinct unit, though they may well have intermingled more had we not been there to worry them. After peering at us for a few minutes longer, they sank below the lilies and vanished from sight. We had no idea how long they had been together before we had come across them, but the party had broken up by the following morning for, paddling down their creek core, we spotted the Maduni family swimming alone. The fact that the other two unidentified groups were not with them added substance to my belief that core areas are exclusively held by one family only. Most probably, the two strange families held cores somewhere nearby in Russel Lake but shared a portion of their home range with the Maduni and Streaky groups.

Far more commonplace than super-groups were the close-knit nuclear families of parents and cubs. And as familiar, though not as extrovert as these cosy social units, were the virgin spinsters and bachelors – young rolling stones who spent their time gathering experience and looking for a regular beat of their own. Two-parent families are scarce in the animal world but even more unusual is a permanent monogamous relationship. It is this long-term fidelity that puts giant otters into a very select group of mammals whose members include jackals, gibbons and beavers. In the three years of study at Lama, none of the four sets of parents changed partners, which suggests that giant otters mate for life. And not only are pairs 'faithful'; neither sex seems to dominate the other. True, the male in the Mazaruni did groom his mate more often than she groomed him, but this does not automatically stamp him as a subordinate. Male status must also be assessed in other ways; for example, males were always the first line of defence. And in the process of marking equality ruled – no one sex consistently landed or sprainted first on the communal sites. In giant otter society it seems safe to say that the sexes are well and truly equal.

Cubs, on the other hand, are most certainly under the thumb of their parents. But they grow quickly and become as bold as their guardians well before their first birthday. By ten months of age, they have learnt to periscope and begin sprainting with their parents on the family's latrines. This *rite de passage* may cause a stir of interest in a passing transient which then follows the family discreetly until the cubs are about to leave home. The apron strings are actually cut when the cubs reach sexual maturity at two years of age. But from what I could record of Dot and Scratch of the Streaky family, they remain within the security of the family home for at least a few months longer. Mister and Missus and their new cub avoided the immediate area during this hardening-off period. It would have been interesting to see when Dot and Scratch did eventually make the big break and leave their parents' territory for good but unfortunately time caught up with us.

Courtship in the giant otter still remains a mystery. No-one has ever witnessed direct courtship, though Shadow's behaviour towards Spotted Dick's cubs could certainly be construed as a form of preliminary courtship. The reason for this long period of follow-the-family was probably as much to sound out the quality of the cubs as potential mates as to put in an early order for one of them. The fact that I never discovered Shadow's sex leaves open the question as to which sex makes the running in courtship. The equality that operated in cub-rearing may also extend to finding a mate. If this is so, then the river wolf is still several steps ahead of the human race, where the males do most of the chatting up.

Once the ball starts rolling, courtship in the giant otter is probably a lengthy affair because in a lifelong marriage each partner must be sure of the other's quality. Neither party can be too careful in a relationship that is for keeps. Nonchalance in choosing could mean you end up a socio-biological failure, producing few cubs or none at all. On the other hand, where the pair-bond is only transitory, as in the European otter for example, courtship is brief and is initiated by the male; a European dog otter is unsure of the female's quality and fidelity, so he must minimise the time and energy spent in courtship. In his sort of social milieu far more profit (offspring) is gained by mating with as many land-owning females as possible. Giant otter Romeos and Juliets are likely to be far more discriminating, sizing each other up for strength, health, compatibility, hunting prowess and possibly even smell. It is not at all improbable that certain types of smell are more adept at defending territories and conveying status than others. In human terms, the idea of choosing your mate for his after-shave or her perfume sounds inordinately fickle, but for any scent-marking mammal, odours are probably a very important facet of its attractiveness: a case of good scents making good sense.

Apart from ample space in which to court, feed and raise a family, it was plain that giant otters needed plenty of shady, unfloodable land. Cover was not all that important at Lama because there was not a great deal of human disturbance, unlike the otter's situation in Britain where intense recreational disturbance sharpens the need for a dense cover of low bushes. In fact, otters as a whole are particularly vulnerable animals. Eliminate any one of their environmental requirements – a large undisturbed home range, plenty of dry, shady land and the right kind of food – and whole populations can plummet in a matter of a few years. This is especially true of the giant otter; it is not only a slow breeder but, being rare to begin with, its small litters and long cub-development mean that it makes heavy weather of staging a comeback. Sometimes interference can take a more subtle form by preventing the breeding pair from reproducing or by causing a total family break-up. Nicole Duplaix once witnessed a pitiful chain of events when a hunter snatched a pair of young cubs from their parents. The adults seemed to lose all interest in life and soon afterwards lost their territory to a neighbouring family group. A few weeks later, the female disappeared without trace, leaving her mate to eke out a subordinate existence in a tiny enclave.

We often wondered if Nibs would succeed in establishing his own territory in Russel Lake. By now we were confident of his ability to feed himself, but as the time for his release approached, we became increasingly pre-occupied with the dangers he would face. Apart from finding a home of his own, our worst fear was that his friendship with humans might make him more vulnerable to hunters than any of his wild peers.

Besides these practical worries, there was also an emotional side to losing our mischievous foster son. The small spark of relief I felt in knowing that I would no longer have to spend hours each day procuring Nibs' two-and-a-half pounds of fresh fish and cleaning up his spraint was far outweighed by the sadness of losing this lithe bundle of energetic mischief. In the twelve months he had been with us he had become part of our lives: a little pest at times, but what he lacked in obedience he more than made up for in charm. Nibs had brightened our stay at Lama, bringing comic relief to situations that would otherwise have been dull. It seemed almost impossible to think of life without him, of showering in the morning without having our toes nibbled, of never playing 'wobble-belly' again. But returning him to the wild was best for Nibs and we knew we had no option.

I measured him for the last time. Earlier attempts at doing so with a tape were farcical, one of those procedures that sound simple but are hard to put into practice. He obviously thought it was a new game and squirmed every way to try and get the tape between his teeth. After several such trials and some ridiculous figures that said he was shrinking, I resorted to photographing him against a metre rule. At two feet eleven inches, he was now pretty well adult size.

It was not until the final day that we could bring ourselves to say goodbye to our now almost fully grown otter. We released him about two in the afternoon, on the Russel Lake side of the dyke, halfway between Lama and the area where we had seen a wild Guyana otter a few days before. Nibs did not seem to have an inkling of what was ahead; he swam alongside the boat on his long leash, occasionally impatient of its limitations but otherwise quite content. We had deliberated long and hard on whether to feed Nibs well or to give him no food at all before letting him go. In the first instance he might not feel inclined to leave, and we wanted him at least to try. But equally, to send him out into the hard, cruel world without a full belly seemed totally inhumane. That last meal might tide him over the most critical period: his first few days of freedom. It might just be the difference between life and death. In the end, we settled for a good old British compromise, though I must admit we erred rather on the side of too much as against too little feeding.

Keith wanted to cover the action for the end sequence of the film, so he had me and Nibs paddle and swim our way down the canal a few times while he filmed from the bank. It was an awful time. I never felt less like being filmed; it seemed a totally in-sensitive thing to do, to spend time posing for different angles, close-ups, medium- and wide-angle shots of something that emotionally meant so much to us. But it was all part of the job and there was nothing for it but to press on and get the shooting over as quickly as possible.

Once we were finished, I called Nibs back into the boat and rolled him onto his back to distract him with a game of wobble-belly while at the same time unfastening the collar around his neck. Keith was still filming; this last shot was the most important. I pulled the collar off so that Nibs was now completely free, but he didn't seem to know it. He still stayed on his back in the bottom of the boat, expecting more wobble-belly play. I looked at Keith helplessly (we were getting low on film, and had expected Nibs to quit the boat immediately), then bent forward to continue the game.

At that moment the harsh bark of a giant otter resounded across Russel Lake, followed swiftly by a chorus of 'wah-wahs' and sneeze-grunts. It was so unexpected that both Keith and I were stunned. The suddenness of the sound completely unnerved Nibs and, as always when he was afraid, he made for the safety of the water. Nibs disappeared over the side of the boat almost without a splash. The whining calls continued for about a minute, moving further away in the direction of Lama but, although we knew for certain they were giants, we never caught even a glimpse of the creatures. It was the same with Nibs. From the time he dropped over the side, he simply disappeared. What else we expected from him I am not really sure, but I think I had it in my mind's eye that he would swim around the boat for a while so we could in some way reassure ourselves that Nibs really was ready for his change to an independent life. The fact that he had fled so precipitately left me with the feeling that I had somehow been cheated of my last lingering look at the creature that had for so long been our friend.

172

Remembering Nibs' first, unintentional release above Kaieteur Falls, we settled ourselves down for a long wait. We would stay in the area until the sun went down at seven in the evening and then motor slowly home. The afternoon slipped by, giving us plenty of time to dwell on the doubts and uncertainties of releasing Nibs. We knew it was for the best, but the idea of him out there in Russel Lake alone pulled at our heartstrings. We had just decided that an hour before we left we would try calling for Nibs when, not far from the boat, we saw the bankside grasses shaking with the characteristic movement that speaks of an animal moving through them. We both brightened immediately and I looked at my watch: two and a half hours since he had jumped ship and it looked as if Nibs was on his way back to us. We watched the unseen animal approaching with idiot smiles on our faces, so pleased to have our otter back with us. Then, as it drew level with our canoe, there was a 'plop!' as something dropped into the water and out from between the grass stems swam . . . a striped grebe!

I suppose we should have been relieved that our plan had worked and that Nibs really had decided to go off alone, but this little episode made us more depressed than ever. When Keith asked me to sit at the end of the boat while he took some sad shots of me against the setting sun, he knew I would not have to act – we both felt utterly miserable. The sun finally went below the horizon and there was still no sign of Nibs. We waited another half an hour and set off for home. Nibs was gone.

17 Going Home

(Keith)

By the time we arrived back at Lama we both felt a little better. Over our supper we managed to persuade ourselves that, despite the sorrow we naturally felt at Nibs' release, it was far better for him to be out there on Russel Lake than facing an uncertain future with us in Georgetown. Even the best option, staying with the Chin family, was in doubt. It had been several months since they had made their offer to foster Nibs, and we had heard nothing since. Had he refused to leave, there was no guarantee that the Chins would still take him. We were still talking about Nibs – we had talked of nothing else – when the lights went out. Of course! It was Thursday and lights-out was at 8.30, a token gesture by the EDWC towards decreasing the amount of imported oil. We fumbled around in the inky blackness, looking in rucksack pockets, and eventually finished our meal by torchlight.

Feelings of regret began to sneak back into my mind. Bats swooped low over the waters, and the eyes of several caiman reflected back the torch-beam. Pale moonlight was flooding across Russel Lake, suffusing the whole broad panorama with a sickly, silver-grey light. The place looked so malevolent, so huge and implacable, that the thought of Nibs lying on some bankside completely alone filled me with melancholy.

Liz led the way upstairs, and she had just reached the top landing when she suddenly stopped with a gasp of surprise. I pushed past her and there, spotlit in the torch-beam, curled up fast asleep against the bathroom door, was the long, lithe shape of an otter. Nibs had come home!

The human mind is a strange thing. I was no sooner overjoyed at Nibs' return than I became terribly anxious at the problem his presence now posed. The little otter was utterly dishevelled and completely exhausted. It must have taken a herculean effort for him to travel so far so fast. And how had he remembered the route, especially in darkness? We washed him, fed him as best we could and put him to bed on top of two pairs of our old jeans. He was asleep at once, totally unaware of the worries he had once again thrust upon us. We had five days back in Georgetown before we left for England. Could we do anything in this brief period to give Nibs a secure and happy future?

Next morning, we loaded our baggage onto the speedboat for the last time and said our goodbyes to Lama, to Pappy and to Katina. Seven hours later we were again smuggling Nibs past the concièrge in Georgetown's Park Hotel. As soon as he was safely deposited in the flooded bathroom with his toys, I 'phoned through to the Chins' house. It was a terribly tense time: Neville Chin's answer would mean the difference between a pleasant life for Nibs or a terribly restricted existence, either in Georgetown Zoo or in the cold, alien climate of Britain. The 'phone seemed to ring forever before it was answered. When someone did lift the receiver, it was the youngest member of the family, Judy. She told us that the rest of the family were out but would be back in about an hour. In the next breath she was asking excitedly about Nibs – was he still with us? Judy was delighted with my answer. Could she come over and see him again? Of course, I agreed, but would she ask her parents to 'phone us as soon as they returned? It was very important.

I put the 'phone down and we both sat on the edge of the bed to wait out the longest hour of our lives. We could not help talking over the possibilities, and we grabbed wildly at any hint, no matter how small, that might indicate how Neville and Pauleen Chin felt about their offer to take on an otter foster child. Judy, I remembered, had said she would love to come over. That meant that the children were still very keen on seeing Nibs. But why, we puzzled, if they were really going to take Nibs, had she not told us to come over and to bring Nibs along? On the other hand . . . It was foolish, of course, as there was no way of knowing at that time what path events might take. We simply had to suffer through the sixty minutes until the 'phone rang. Fortunately, we did not have to undergo the full one hour's torture. The 'phone jangled into life after half an hour. I hesitated a moment, then picked it up and Neville Chin asked, 'When does our new son arrive?'

We hired a car and made the half-hour drive to Ogle Plantation in twenty minutes. The Chins' house was set in a complex of eight buildings completely surrounded by a high wire fence which divided the compound from the encircling cane fields and their feeder canals. The whole Chin family were assembled on the lawn in a welcoming committee as we arrived, and they insisted on taking Nibs on a guided tour of his new home immediately. The ground-floor loo had been specially set out for him; Pauleen told us that she and the children had been preparing it 'just in case'

175

ever since they had first offered to care for Nibs. The shower was already full of water, there was a sleeping mat on the floor, a pile of small fish in the corner and toys galore. Nibs took to all this splendour as to the manner born. He jumped straight into the shower pool, sniffed around the sleeping mat before rolling on it to dry himself, ate a few fish and finally, with his truest mark of acceptance, sprainted under the sink.

Having seen the interior decor, Nibs was led outside to inspect the garden. He had the run of a long, green lawn at one end of which was his own exclusive outdoor swimming pool. This was an old forty-gallon oil-drum, full to brimming with creek water and with a timber gangway leading up to its edge. Hardly deigning to look at the assembled humans, all eagerly awaiting his reaction, Nibs scampered up the walkway and plunged deeply into the artificial pool. He emerged several seconds later with a large live patwa – a house-warming present caught by Steven – dangling between his jaws. To Nibs, the Chins' home must have seemed like Paradise, but there was better to come. At the bottom of the garden was a narrow trench which linked ultimately with the main canal that ran southwards to join with Russel Lake some three miles distant. In this trench were hundreds of small fish – an otter delight!

In the course of his explorations, Nibs also met some of his new neighbours. George, the Chins' Doberman guard-dog, was plainly astonished by this new addition to the family, but seeing how the humans adored the newcomer, he went out of his way to be nice. Nibs' experiences with the dogs at Lama seemed to have removed his fear of dogs, no matter what their size. Although George towered above him, he boldly ran for the dog's legs and proceeded to play-nibble the dog's toes. The Doberman was completely nonplussed by this strange animal's behaviour but, wanting to respect its owner's wishes, it did not bite back. Instead, it proceeded to prance backwards, lifting each foot in turn as Nibs attacked and looking for all the world like a canine version of Vienna's Spanish horses!

Neville Chin bred fighting cocks too, and Nibs found these just as chaseable as George. But he met his match in the form of a small bantam hen with her six chicks. Nibs charged forward eagerly, but the little mother puffed up her feathers and, to his great surprise, responded to his attack with a charge of her own. Nibs skidded to a halt in the face of this determined resistance and he was soon flying round the garden hotly pursued by his clucking victor. He escaped only by plunging ignominiously into the trench water. However, like all otters, Nibs knew how to turn even defeat to his own advantage, and he had soon forgotten his shameful thrashing in the excitement of chasing the numerous trench fish. By the time we were ready to leave he seemed completely at home. When we stepped into our car Nibs had vanished; his presence was confirmed only by the sound of fish being eaten beneath the trench bridge.

During the next few days we visited the Chins' house several times to make sure that Nibs was really settling in as well as it appeared. We need not have worried: with all the stimulation from the three children and their parents I do not think Nibs ever had the time to miss us. And he had other friends, too. George the Doberman had gradually been won over during the first day, and had even introduced Nibs to his lady-love, a flea-bitten mongrel from a neighbouring house. These three began foraging together around the compound in the early morning. We were quite concerned by this turn of events: our big worry was that this friendship would eventually 'doggify' Nibs. As it fell out, the opposite was true. We arrived one morning to find not one but three water dogs swimming in the canal: Nibs, George and his girl-friend were all paddling happily through the water. Pauleen had watched it all and told us that every time their foraging had brought them to a canal Nibs had leaped blithely in to swim and dive. The two dogs both sat patiently at the water's edge, waiting for their friend to finish. But Nibs was not satisfied with swimming alone. He kept stopping and wheeping at the two dogs, apparently giving the otter equivalent of 'Come on in, the water's lovely'. Eventually, the more adventurous female had taken the plunge and, after a few minutes hesitation, George had followed suit. They really did seem to enjoy the feel of the water, although Nibs made his superiority in this department very clear by swimming rings round his two friends.

Nibs' ability to cope with new conditions made us considerably less worried about his future. Some otters, we knew, identify so strongly with their human companions that they suffer severe psychological trauma when separated. But we could not help admitting that, although we were pleased by his adaptability, we also felt just a little hurt by his fickle nature. After all that time, all those games, the fishing expeditions, the fire ants. . . . Still, we knew it was for the best and far, far preferable to him pining.

We went back for a flying visit on the morning of our flight to Britain. We wanted to say a quick goodbye and then be on our way. But when we arrived it was to find that Nibs had already flown the coop. Over the days he had gradually been exploring the waterways, moving further and further afield. The previous

177

morning Nibs had finally found the main canal and had swum steadily up it. He had been seen entering Russel Lake around mid-day by two cane-cutters. Nibs had not returned that night, and although the Chin family had spent most of the small hours walking the canals, calling out his name, he had not responded to their shouts. Just before we arrived, another estate worker had told Pauleen that he had observed Nibs (whom he had seen several times in the compound) feeding on a large patwa in the shallows of Russel Lake. It seemed that our little otter had at last decided that he could fend for himself. The call of the open water had proved stronger than human companionship and the comforts of a civilised existence. It was the best news we could have had for our departure.

Nibs had chosen a good spot to set up home. Russel Lake is absolutely essential to Georgetown's water supply and it is likely to remain out-of-bounds to humans for a very long time to come. While other habitats yield to Man's frightening population increase Russel Lake at least will remain inviolate. That means safety for those animals lucky enough to establish a home on its broad waters. If Nibs can do this and find a place among his own kind, then his future, like that of Russel Lake's giant otters, seems secure.